Advanced Structured Materials

Volume 10

Series Editors

Prof. Dr. Andreas Öchsner, Technical University of Malaysia, Skudai, Johor, Malaysia

Prof. Dr. Holm Altenbach, University of Halle-Wittenberg, Halle, Germany

Prof. Dr. Lucas Filipe Martins da Silva, University of Porto, Porto, Portugal

For further volumes:
http://www.springer.com/series/8611

Fernando A. Lasagni · Andrés F. Lasagni
Editors

Fabrication and Characterization in the Micro-Nano Range

New Trends for Two and Three Dimensional Structures

Springer

Editors
Dr. Fernando A. Lasagni
Center for Advanced Aerospace Technologies
Andalusian Foundation for Aerospace Development
Seville
Spain
e-mail: flasagni@catec.aero

and

Metallurgy and Materials Engineering Group
Department of Materials and Mechanical Engineering
Seville University
Seville
Spain
e-mail: flasagni@us.es

Dr. Andrés F. Lasagni
Fraunhofer Institute for Material and Beam Technology
Dresden
Germany
e-mail: andres-fabian.lasagni@iws.fraunhofer.de

ISSN 1869-8433 e-ISSN 1869-8441

ISBN 978-3-642-17781-1 e-ISBN 978-3-642-17782-8

DOI 10.1007/978-3-642-17782-8

Springer Heidelberg Dordrecht London New York

Cover design: WMXDesign GmbH, Heidelberg

Printed on acid-free paper

Springer is part of Springer Science+Business Media (www.springer.com)

To our mother, Renata.

In loving memory of our father, Norberto.

We would like to dedicate this book to our beloved Inma (F.L.) and Nicole (A.L.), whose understanding, support, and encouragement made the completion of this book possible. This work is also dedicated to my son Marco (F.L.) and nephew (A.L.).

Andrés Lasagni and Fernando Lasagni

Preface

Fabrication of two and three dimensional structures in the micro- and nanorange allows a new degree of freedom to the design of materials by tailoring desired material properties and in this way obtaining a superior functionality. Such complex designs are only possible by using novel fabrication techniques with high resolution, even in the nanoscale range. During the last years, different methods have emerged, allowing local modification of materials properties in two or three-dimensions as well as the rational design of complex three dimensional architectures. In this way, it has been possible to improve physical, chemical and biological properties of all classes of materials including ceramics, metals and polymers. Moreover, it was observed that even internal microstructural inhomogenities such as grain boundaries and porosity can further enhance their performance.

However, the future progress for the development of new advanced materials depends not only on the ability to fabricate them, but also on understanding underlying nanoscale and interface effects as well as through a truly three dimensional characterization of their structure. In consequence, adequate and high detail characterization of these structures is necessary. This is nowadays possible due to new developments in tomography techniques which allow their complete three-dimensional reconstruction in all ordering scales. These tools have been also applied recently to investigate time-dependent properties such deformation mechanisms of structural materials.

This book comprises a series of monographs elaborated by contemporary experts. It outlines the underlying of scientific principles for a broad sample of both fabrication and characterization techniques from the micro to the nanoscale. The book is divided into two sections, a first part describing two and three dimensional fabrication techniques, and a second one for the latest characterization technique at the same resolution scales.

The first part of the book consists of five chapters. Chapter 1 discusses the possibilities of Laser Interference Patterning for the rapid fabrication of periodic patterns, with a well-defined long range order from the submicrometer level up to micrometers, on macroscopic areas. Chapter 2 describes different laser assisted microfabrication techniques as well as typical applications of laser micro machining including drilling, cutting, and marking. Chapter 3 deals with the fabrication of nanostructured materials with enhanced optical properties using different methods such as nanoimprint lithography and nanosphere self-assembly. In addition, also the characterization of the optical properties is introduced. Chapter 4 focuses on self-organized pattern formation by ion beam sputtering technique. A general description of the method and experimental observations are presented, showing the complexity of the processes involved. The great potential of the technique is presented as well. Finally, Chap. 5 describes the use of Additive Manufacturing Technologies for fabrication of cellular architectures with enhanced mechanical properties. In particular, an overview on two-photon polymerization method is given.

The second section of the book has four chapters. Chapters 6 and 7 deal with different X-ray tomographic methods including micro-focus X-ray computed tomography as well as submicron tomography using high energy synchrotron radiation. In the last case, the application of the optical X-ray phase of the transmitted beam for increasing the sensitivity of the method is also described. Different examples for the 3D characterization of structures as well as defect analysis are given. Chapter 8 introduces the method of Focused Ion Beam Tomography, which allows the nano-characterization of three-dimensional geometries by reconstructing the information provided from two-dimensional cuts (slices) imaged either with electrons, X-rays or ions detectors. Finally, Chap. 9 describes the method of Atom Probe Tomography, which is able to map out the 3D distribution of chemical species in a material at the atomic scale. In addition, new alternatives that permit the analysis of semi-conductors or oxides are introduced.

The methods here described illustrate that there are often various approaches for both fabrication and characterization of micro and nanostructures using in many cases equipment that is commercially available or can be readily assembled. The book is indented to serve as a resource text for undergraduate and graduate students, as well as for researchers from materials science and engineering.

We would like to thank several colleagues for their kind assistance in proofreading at various stages. We would like to especially thank all the contributing authors: Blavette, D., Borbély, A., Cloetens, C., Cornejo, M., Deconihout, B., Franke, V., Frost, F., Harrer, B., Kastner, J., Klotzbach, U., Luxner, M., Maire, E., Menand, A., Mücklich, F., Olaizola, S., Panzner, M., Pérez, N., Pettermann, H., Rauschenbach, B., Requena, G., Rodríguez, A., Soldera, F., Stampfl, J., Völlner, J., Vurpillot, F., and Ziberi, B., as well as Castro M. (Springer) for their support during the preparation of the book. This book has been possible because of all of them.

Dresden, November 2010 Dr. Andrés F. Lasagni
Seville, November 2010 Dr. Fernando A. Lasagni

Contents

Exploring the Possibilities of Laser Interference Patterning for the Rapid Fabrication of Periodic Arrays on Macroscopic Areas

Andrés Fabián Lasagni

Abstract Surface patterning engineering techniques are essential to fabricate advanced topographies that can be use to modulate macroscopic properties on different materials. Particularly, Laser Interference methods enable fabrication of repetitive periodic arrays and microstructures by irradiation of the sample surface with coherent beams of light. Depending on the used laser source, different methods have emerged in the last years including Laser Interference Lithography and Direct Laser Interference Patterning. A detailed description of these techniques is presented in this chapter. In addition, several examples including fabrication of micro and sub-micrometer patterns on photoresists, conducting polymers and carbon nanotubes are described.

1 Introduction

Fabrication of periodic surfaces with micron and submicron features is a rapidly growing research field with applications in several technological areas. Such structures on the surface of metals, semiconductors, dielectrics or polymers can generate new material properties with very special electrical, mechanical or chemical characteristics. Depending on the specific material parameters and the morphology of the structures, new devices like bio-sensors [1], antifraud features, microfluidic devices [2–5], templates for biological applications [6] as well as photonic structures [7] can be realized. Furthermore, surface textures can be used

---

A. F. Lasagni (✉)
Fraunhofer Institute for Material and Beam Technology (IWS), Winterbergstr 28, 01277 Dresden, Germany
e-mail: andres-fabian.lasagni@iws.fraunhofer.de

F. A. Lasagni and A. F. Lasagni (eds.), *Fabrication and Characterization in the Micro-Nano Range*, Advanced Structured Materials, 10,
DOI: 10.1007/978-3-642-17782-8_1,

to improve tribological properties of special tools [8], for the reduction of reflection losses [9] or as decoration elements for the refinement of precious goods.

Apart from that, photon based fabrication methods offer several advantages due to their remote and thus contactless operation, their flexibility during materials processing, as well as their precise energy deposition. At the micro scale, laser direct writing and micro stereolithography techniques are being utilized to pattern and process several materials with features in the range of 1–100 μm [10, 11]. In addition, in some cases an imaging setup makes it possible in one-single step operation to reproduce a specific pattern by local ablation, photopolymerization or surface modification processes. However, these methods require masks in order to obtain a specific geometry or pattern shape over a substrate, or can be quite time consuming. In order to solve this problem, techniques involving large-area maskless lithography for rapid fabrication of two and three dimensional (2 and 3D) structures are required.

Laser Interference Lithography (LIL) has been utilized in the past for large area fabrication of planar periodic structures [12, 13]. To generate an interference pattern, N collimated and coherent laser beams must be overlapped over the material surface. One of the principal advantages of LIL is that no masks are needed and the shape and dimension of the interference patterns can be just adjusted by controlling the number of laser beams as well as their geometrical configuration (polar and azimuthal angles of each laser beam) [13–15]. In LIL, a photoresist is normally exposure with UV radiation. Then, the more exposed and less exposed parts have different solubilities in the developer, leading to a porous structure [16]. Finally, the irradiated surface is etched, eroding the positions without photoresist obtaining the final structure. On the other hand, fabrication of 2D and 3D structures requires more than two coherent laser beams. However, increasing the complexity of the experimental setup also increases the possibility of introducing defects in the periodic structure [17], as will be later discussed.

Recently, high power laser systems have been used to directly process the surfaces of different materials, without employing a photoresist [18–21]. In this way, based on photo-thermal and/or photo-chemical interactions, the surfaces of the materials could be directly and locally modified at the potions corresponding to interference maxima. Therefore, the name DLIP, or DLIL was introduced [22, 23]. Moreover, since also several metallurgical effects can be induced (especially in metals and ceramics), the method has also been called Laser Interference Metallurgy (LIMET) [18, 24].

In this chapter, different aspects of the "Laser Interference" based methods, including LIL, DLIP and LIMET will be introduced. In the first part of the chapter different approaches for the fabrication of 2 and 3D structures using two- and multi-beam interference patterns will be described. Particularly, a model for the calculation of interference patterns with N laser beams, discussing the effects of individual beam phase as well as incident angles on line-defects is presented. After that, several examples of pattern fabrication on photoresists, conducting polymers, carbon nanotubes (CNT) and other materials are discussed showing the potential of the method.

2 Calculation of Multi-beam Interference Patterns

To calculate the intensity distribution of an interference pattern, it is necessary to add together all overlapping laser beams as follows:

$$E = \sum_{j=1}^{N} E_j = \sum_{j=1}^{N} E_{j0}\, e^{-ik \sin \alpha_j \left(x \cos \beta_j - y \sin \beta_j\right) + \psi_j}, \tag{1}$$

where E_j are the amplitudes of electric field of each j-beam; α_j and βj which are the angles of the beams with respect to the vertical (polar angle) and the horizontal axis (azimuthal angle) (Fig. 1) of the interference-plane, respectively; ψ_i is the initial phase, and k the wave number:

$$k = \frac{2\pi}{\lambda}, \tag{2}$$

with λ denoting wavelength.

Then, the total intensity of the interference pattern can be calculated as

$$I = \frac{c\varepsilon_0}{2}|E|^2, \tag{3}$$

where c is the speed of light and ε_0 the permittivity of free space. Using the last Eq. 3, the interference pattern of N-beams can be easily calculated [14, 25].

For two laser beams configuration, a 1D line-like interference pattern is obtained (Fig. 2a, with $\alpha_1 = \alpha_2 = 20.34°$, $\beta_1 = 0$, $\beta_2 = \pi/2$, $\lambda = 355$ nm, and $\psi_1 = \psi_2 = 0$). In this case, independently if the polar angles α_j are the same (with $j = 1, 2$), no line defects are observed as well as no changes in the intensity profile. In addition, any offset of the pattern is observed when maintaining the total intercepting angle between the beams constant ($\alpha_1 + \alpha_2 = C$) and the periodic distance between two intensity maxima (period: Λ) remains unchanged. From the experimental point of view, this situation is common in the case that the sample is not placed exactly at the position of the $z = 0$ plane showing a tilted angle θ (see Fig. 1) and represents an important advantage when fabricating the periodic arrays. On the other hand, a shift of the initial phase (ψ_j) induces an offset of the whole pattern but no variations of the intensity are observed. The maximal offset is obtained for a phase difference of 180° (π) and corresponds to half of the period ($\Lambda/2$) moving from right to left as shown in Fig. 2b (with $\alpha_1 = \alpha_2 = 20.34°$, $\beta_1 = 0$, $\beta_2 = \pi/2$, $\lambda = 355$ nm, and $\psi_1 = \pi$; $\psi_2 = 0$) [26].

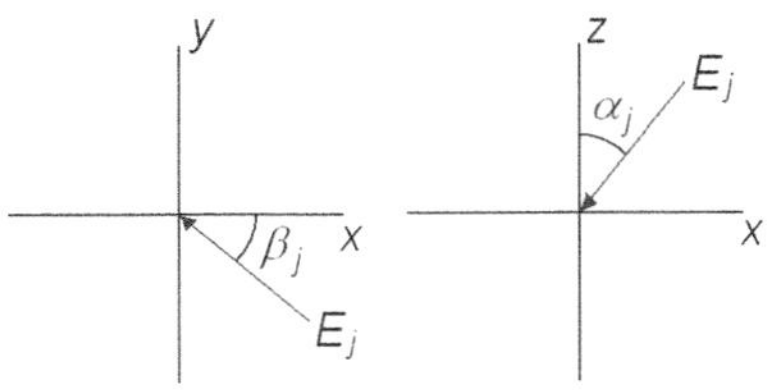

Fig. 1 Representation of an electro-magnetic wave in 3D. The sample is located at the position $z = 0$ parallel to the plane x, y (interference plane)

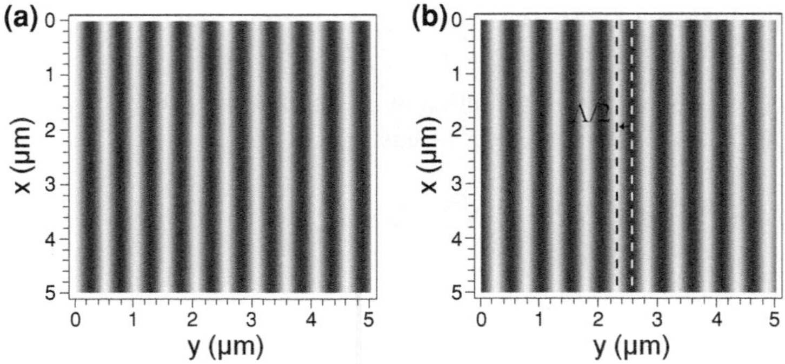

Fig. 2 Calculated intensity distribution of the two-beam interference patterns with $\alpha_1 = \alpha_2 = 20.34°$, $\beta_1 = 0$, $\beta_2 = \pi/2$, $\lambda = 355$ nm and **a** $\psi_1 = \psi_2 = 0$; **b** $\psi_1 = \pi$, $\psi_2 = 0$. The offset between both patterns is $\Lambda/2$ moving from right to left when increasing the phase ψ_1. Reprinted figure with permission from Lasagni et al. [26], Copyright (2010) by WILEY–VCH Verlag GmbH & Co (Editor Jörn Ritterbusch)

In the case of four laser beam interference, a very different situation is observed. If the beams are incident on the sample having the same polar angles α_j or satisfying the condition $\alpha_1 = \alpha_3 \neq \alpha_2 = \alpha_4$ with $\beta_1 = 0$, $\beta_2 = \pi/2$, $\beta_3 = \pi$, $\beta_4 = 3/2\ \pi$, a uniform intensity distribution is obtained (Fig. 3a, with $\alpha_1 = \alpha_2 = \alpha_3 = \alpha_4 = 20.34°$, $\lambda = 355$ nm, and $\psi_1 = \psi_2 = \psi_3 = \psi_4 = 0$). On the contrary, for an asymmetric configuration which is not satisfied by the conditions previously mentioned, periodic or quasi periodic linear defects appear in the pattern [17]. This effect is shown in Fig. 3b for the condition $\alpha_1 = \alpha_2 = \alpha_3 = 20.34°$, $\alpha_4 = 23.34°$, $\beta_1 = 0$, $\beta_2 = \pi/2$, $\beta_3 = \pi$, $\beta_4 = 3/2\ \pi$; $\lambda = 355$ nm, and $\psi_1 = \psi_2 = \psi_3 = \psi_4 = 0$. The line defects consist on a periodic variation of laser intensity perpendicular to the laser beam with the asymmetric angle (in this case the beam $j = 4$, with a 3° difference compared to beams 1–3). Apart from that, for the symmetrical configuration case ($\alpha_1 = \alpha_2 = \alpha_3 = \alpha_4$) a phase change in one of the beams also produces a change in the shape of the intensity distribution of the interference pattern as shown in Fig. 3c with $\alpha_1 = \alpha_2 = \alpha_3 = \alpha_4 = 20.34°$, $\beta_1 = 0$, $\beta_2 = \pi/2$, $\beta_3 = \pi$, $\beta_4 = 3/2\pi$, $\lambda = 355$ nm, and $\psi_1 = \pi$, $\psi_2 = \psi_3 = \psi_4 = 0$. On the other hand, for a non symmetrical configuration, variation of the phase ψ_j causes no changes in intensity profile but results in an offset of the whole pattern [17]. In consequence, for a proper defect-free periodic structure fabrication using a four laser beam setup, a perfect control of the initial phase, azimuthal angles as well as a perfect positioning of the sample at the plane $z = 0$ is necessary. In contrast, we have seen that when using two laser beams only a change of the pattern offset is observed when varying the initial phase, remaining both shape and intensity of the pattern unchanged. The last shows important advantages compared to the four laser beam configuration and can be used for the fabrication of 2 and 3D structures by firstly irradiating the sample with a line-like pattern, then rotating the sample a defined angle (parallel either to the polar or azimuthal

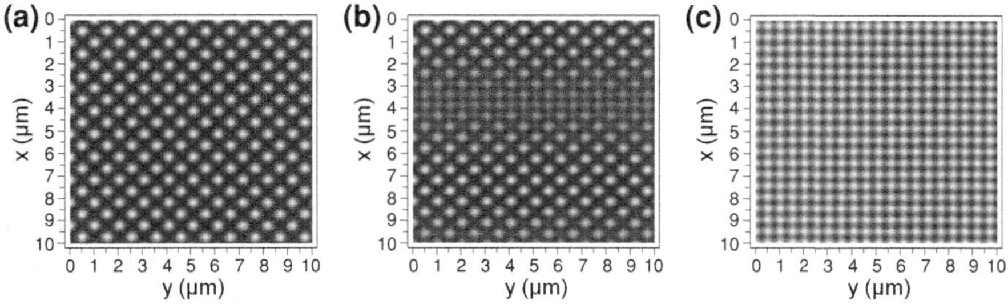

Fig. 3 Calculated intensity distribution of four-beam interference patterns with $\beta_1 = 0$, $\beta_2 = \pi/2$, $\beta_3 = \pi$, $\beta_4 = 3/2\pi$, $\lambda = 355$ nm and **a** $\alpha_1 = \alpha_2 = \alpha_3 = \alpha_4 = 20.34°$, $\psi_1 = \psi_2 = \psi_3 = \psi_4 = 0$; **b** $\alpha_1 = \alpha_2 = \alpha_3 = 20.34°$, $\alpha_4 = 23.34°$; $\psi_1 = \psi_2 = \psi_3 = \psi_4 = 0$; **c** $\alpha_1 = \alpha_2 = \alpha_3 = \alpha_4 = 20.34°$; $\psi_1 = \pi$, $\psi_2 = \psi_3 = \psi_4 = 0$. Note the line defect in (**b**) due to a difference of 3° in α_4. Reprinted figure with permission from Lasagni et al. [26], Copyright (2010) by WILEY–VCH Verlag GmbH & Co (Editor Jörn Ritterbusch)

angles), and finally irradiating the sample for a second time with a line-like pattern using the same or a different periodic distance. These approaches will be described in the next sections.

3 Laser Interference Lithography

This technique is emerging as one of the simplest, fastest, and least expensive method for producing highly ordered 1D, 2D and 3D structures with submicron periodicity over large areas.

LIL has been used to fabricate patterned substrates that are used in a variety of applications including photonic crystal waveguides and field-emission flat-panel displays. In addition, patterned photoresist structures can be used to form templates for nanoparticle assemblies that can be incorporated into nano-photonic devices [27].

The dimensionality of the resultant periodic structure is determined by the number of beams involved. Interference among $N < 4$ collimated, coherent beams produces an $N - 1$ dimensional intensity gratings. Because the multiple beams must be coherent to produce an interference pattern, one laser beam from a visible (e.g. Ar-ion, frequency-doubled Nd:YAG, Nd:YVO4) or UV laser (e.g., frequency-tripled Nd:YAG, He-Cd, KrF) is typically divided into multiple beams [28].

In the process of LIL a small number of coherent optical beams incident from different directions are utilized to produce an interference pattern whose intensity distribution is recorded in the photosensitive layer and is later transferred (developed) by thermal and chemical processes. Such in conventional optical lithography, two different types of optical sensitive materials are utilized: *positive* and *negative photoresist* [29]. In a positive photoresist, the regions in the material that are overexposed to light are chemically changed and become soluble to the developer. On the other hand, the underexposed regions, which are exposed to light intensities below the threshold intensity, remain insoluble to the solvent and form the resultant structure. In a negative resist, the regions in the material that are overexposed to light are chemically changed and become difficult to dissolve. Therefore, the material in the overexposed regions remains intact forming the structure and the solvent removes the material in the underexposed regions. After that, the irradiated samples are etched (chemically, plasma etching, etc.) obtaining the finally structure on the substrate [30].

Two experimental arrangements are widely used for two-beam IL. The first setup utilizes a Lloyd's mirror interferometer as shown in Fig. 4a. This is a simple corner cube arrangement with 90° geometry where the left and right halves of the beam are folded onto each other. This arrangement has been used successfully with sources having a high transverse coherence such as a single-mode TEM Ar-ion laser [31]. Expanding the beam allows a large area overlap, especially when the angle of incidence is ~45° and the sizes of beams incident on the sample and the mirror are equal.

The primary advantage of the Lloyd's-mirror is that the spatial period of the exposed gratings can be easily and continuously varied from many microns down

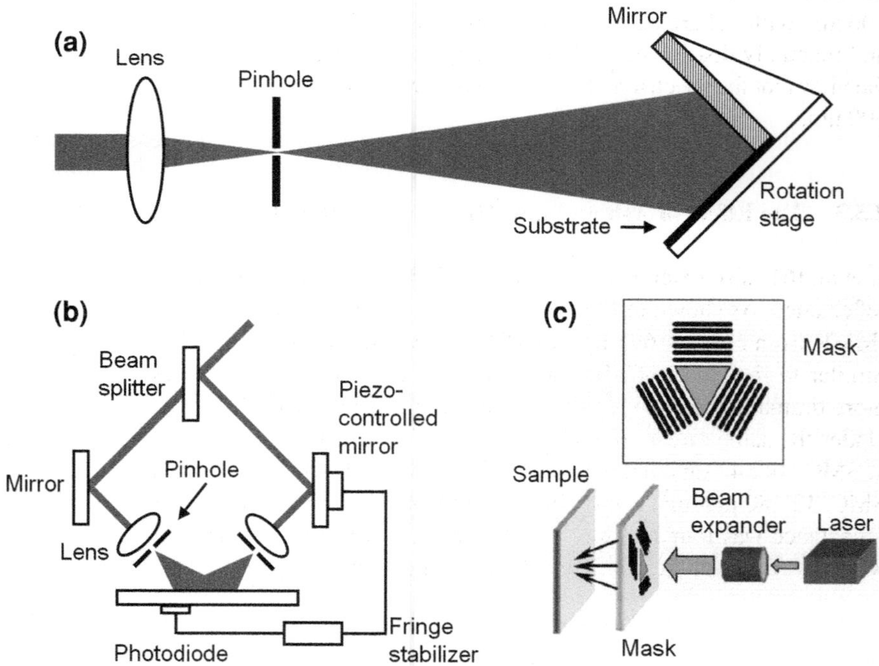

Fig. 4 **a** A Lloyd's mirror interference configuration to fabricate line-like arrays up to ~170 nm; **b** interference setup with a two-beam configuration with beam splitter using a piezo-electrically controlled mirror for phase stabilization; **c** schematic representation of the three-grating diffraction mask and optical setup for creating the hexagonal interference patterns on a photoresist substrate

to ~170 nm simply by rotating the stage without realigning the optical path [32] (Fig. 4a). However, the substrate areas that can be illuminated with any angles of incidence are small. Increasing the pattern area by increasing the mirror size can be expensive because the exacting smoothness and flatness tolerances of the mirrors are difficult to achieve over large areas.

These problems can be minimized by using other arrangements which involve splitting one coherent beam into two ore more sub-beams using beam splitters and then overlapping the beams over the sample (Fig. 4b). Here, the longitudinal coherence (narrow spectral width) is important because the distances from the beam splitter to the image plane can vary for the two beams. Phase errors can accumulate when both beams travel long separate paths and encounter different optics, vibrations, mirror imperfections, spurious scattering and a variety of other deleterious effects. As both beams travel along separate paths, the effects of both vibration and air turbulences can easily disturb them [31]. This situation is common when using ion-argon lasers, where the exposure time ranges from some seconds up to several minutes. For the exposure of large surfaces this problem can be overcome using a fringe-locking system (Fig. 4b). Here, fringe movements detected by a photodiode generate signals that actuate a piezoelectrically driven optical element that stabilize the phase of one of the beams [27, 31].

If pulsed laser systems are utilized, this construction can be done without the fringe-locking systems, avoiding the use of the piezoelectric supported mirrors to compensate low-frequency phase perturbations, keeping the interference fringes fixed during the exposure. This setup will be shown with more detail in Sect. 3.

Recently, there has been resurgence of interest in immersion lithography, the use of a suitable fluid between the imaging lens and the wafer [33, 34]. This allows an improvement in the optical resolution by a factor of the refractive index of the immersion fluid, provided that the lens is suitably modified. At 193 nm, water is a very attractive immersion fluid with a refractive index of 1.44 [35]. The simple addition of a prism to the interferometric arrangement of Fig. 4b allows investigation of immersion IL. In the simplest configuration, a drop of water is placed on the wafer and the wafer and water are raised till the water contacts the uncoated prism and forms a high quality optical component. More elaborate automatic fill and dry systems will be used in manufacturing applications. An example of a 45-nm half-pitch immersion IL pattern is shown in Fig. 5a. After the immersion exposure the sample was baked to activate the chemically amplified resist and developed in a standard basic solution. More details are provided elsewhere [36, 37].

Interference patterns can also be created by passing one laser beam through a multiple grating configuration prior to exposing the photoresist [39]. As shown in Fig. 4c, an expanded incident beam is diffracted by three diffraction gratings incorporated into one monolithic package, and aligned about the normal 120° to each other. This method is stable, fast and easy because it entails using a single monolithic diffraction object. On the other hand, since the interference pattern depends on the period of the gratings and the angle between the gratings, different grating masks are required to change the pattern. In addition, relative low laser intensities must be used in the case of high power pulsed lasers, since the beam can

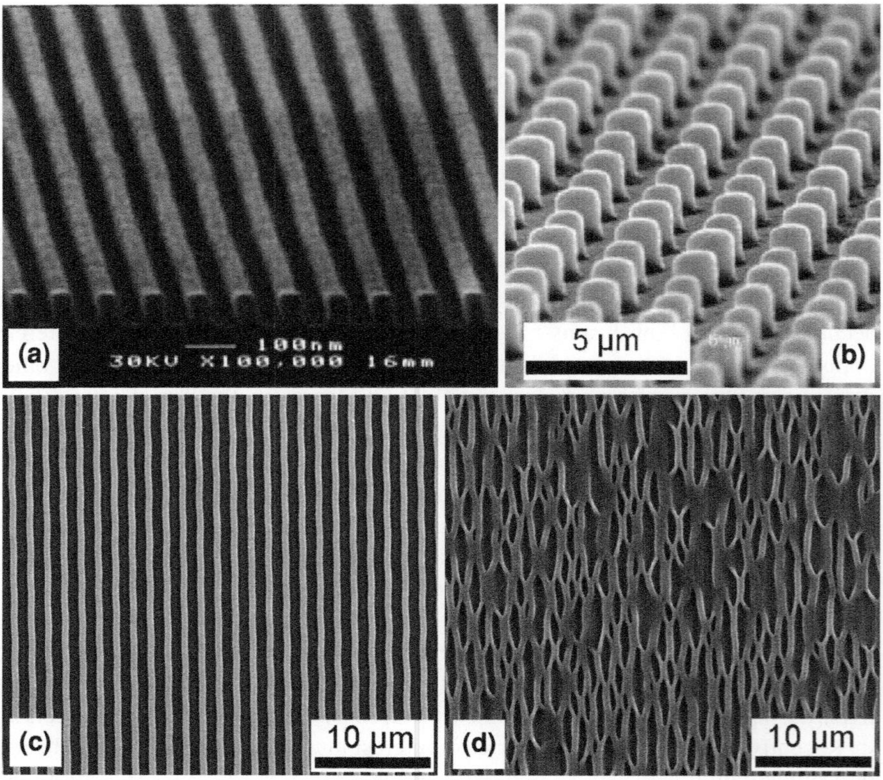

Fig. 5 **a** 45-nm half-pitch pattern in photoresist (resist supplied by AZ Electronic Materials) [37]; **b** hexagonal pillar like pattern fabricated on SU-8 photoresist using a diffractive grating mask (Fig. 4c) [38]. Periodic line-like patterns of photopolymerized SU-8 photoresist with period **c** 1500 and **d** 500 nm and exposure dose of 6.2 mJ/cm^2

damage the gratings. In the example shown in Fig. 4c, a mask with three gratings oriented 120° relative to each another was utilized [38]. Each grating has 2 mm wide and 4 mm long features separated by 2 mm. The photoresist sample is exposed by placing it at the focal point of the diffraction pattern, yielding a hexagonal array of photoresist columns as shown in Fig. 5b [38]. In this way, it is possible to fabricate 2D-photonic structures by forming a polymer template from a photoresist material [38]. As similar setup can be utilized to produce interference patterns with femtosecond lasers and well be described in Sect. 4.

As previously described, using a two-beam experimental setup, different line-like patterns can be reproduced on a photoresist. Using a negative resist (such SU-8), the positions of the material exposed at the interference maxima became solid and after develop of the sample we obtain the final structure. In general, several aspects are critical in order to fabricate stable polymeric structures such as pattern period (Λ), thickness of the resist layer, width of the photopolymerized regions (wall width), pre- and post-baking temperatures and time, and surface tension of developer and/or final wash solvent [40]. However, it can be assumed that line-like patterns are more stable for large periods, small height as well as wide wall-widths. This effect is shown in Fig. 5c and d for two different line-like structures with periods of 1,500 and 500 nm, respectively. In both cases, the exposure dose was 6.2 mJ/cm^2. As it can be observed in Fig. 5d, for the small structure period the strong capillary forces resulting during the final evaporative drying, causing adhesion and collapse of the polymer structures [40]. However, by choosing the correct parameters, it is also possible to fabricate line-like arrays with sub-micron features ($\sim$170 nm). In addition, the development of new resist with higher young modulus will also permit the fabrication of high aspect ratio structures, even with small periodic distances.

Apart from line-like structures, the two-beam configuration setup can be also utilized to fabricated 2D arrays. In the last case, a double exposure process must be performed consisting on irradiation of the sample with a line-like interference pattern followed by a rotation of the sample and a second irradiation of the sample. Moreover, because the periodic structured is cross-linked at the intercepting nodes due to the double exposure process, independently of spatial period, layer thickness or wall width, the structure is stable enough to resist the attractive capillary forces during the rinse and drying steps [41]. Moreover, no line defects appear on the structure since a two-beam configuration is used (see Sect. 1).

Examples of the cross-like structures are shown in Figs. 6 and 7 for 1,500 and 500 nm of spatial period, respectively. By using different rotation angles (30, 60, and 90°) several periodic arrays could be fabricated (Fig. 6). In addition, variation of the exposure dose (laser energy density) permitted to control the width of the photopolymerized lines. For example, for $\Lambda = 500$ nm, a exposure dose of 3.8 mJ/cm^2 results on structures with a wall width of $\sim$160 nm, while for 9.2 mJ/cm^2 the correspondent value is $\sim$435 nm (Fig. 7) [26].

As it can be observed in Fig. 7a, also structures with a small wall width are very stable compared to the line-like patterns (Fig. 5b). On the other hand, if the sample is overexposed, the catalyzer of the photopolymerization (Lewis acid) can diffuse to the non-exposed region inducing cross linking in such regions and consequently closing the holes (see inset in Fig. 7c) [42]. Note that in the inset of Fig. 7a, the hole formation is just starting and are square shaped, while in Fig. 7c the hole sizes are smaller and they are more circular.

In order to qualitatively calculate the shape of the photopolymerized material with the different rotation angles, the following procedure was utilized. Firstly, using Eq. 3, the intensity distribution for a specific two-beam laser configuration was calculated. After that, we calculated the second line-like intensity distribution taking into account a specific sample rotation angle. Finally, both intensity distributions were added and using the Beer–Lambert's law (Eq. 4), the locally amount of photopolymerized was calculated:

$$d = \frac{1}{\alpha} \ln\left(\frac{I}{I_{\text{th}}}\right) \tag{4}$$

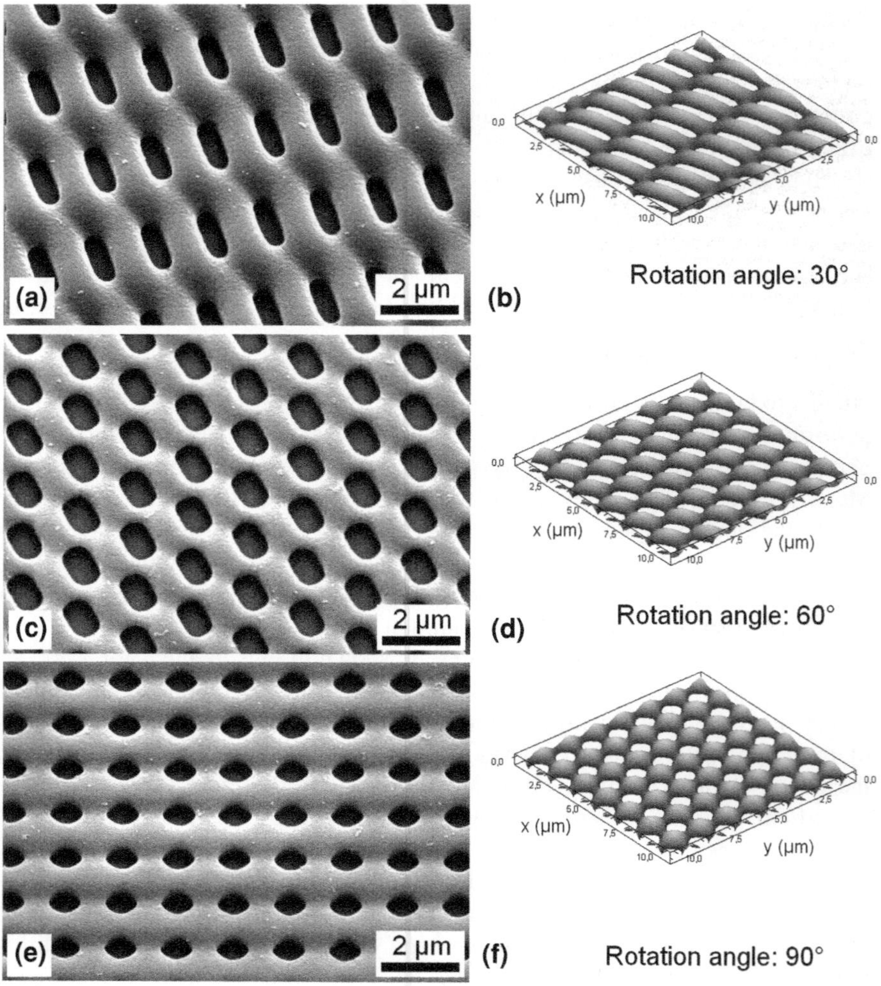

Fig. 6 Periodic two-dimensional patterns of photopolymerized SU-8 photoresist and their correspondent numerical calculation of solidified material. The period Λ of the line-like interference pattern was 1,500 nm and the rotation angles were (**a**, **b**) 30°; (**c**, **d**) 60° and (**e**, **f**) 90°. **a**, **c**, **e** Scanning Electron Micrographs with Tilt angles of 30°. In all case, the exposure dose per irradiation step was 6.4 mJ/cm^2. Reprinted figure with permission from Lasagni et al. [26], Copyright (2010) by WILEY–VCH Verlag GmbH & Co (Editor Jörn Ritterbusch)

where d is the photopolymerized depth at the position x, y; α the absorption coefficient of the photoresist, I the intensity of the total interference pattern at x, y (Eq. 3); and I_{th} is a threshold value at which the material is polymerized.

The calculated surface structure for each fabricated pattern can be observed on the right side of Figs. 6 and 7. The results show a good agreement with the experimental results and permit to qualitatively describe the shape of the fabricated structures as function of experimental configuration and exposure dose [26].

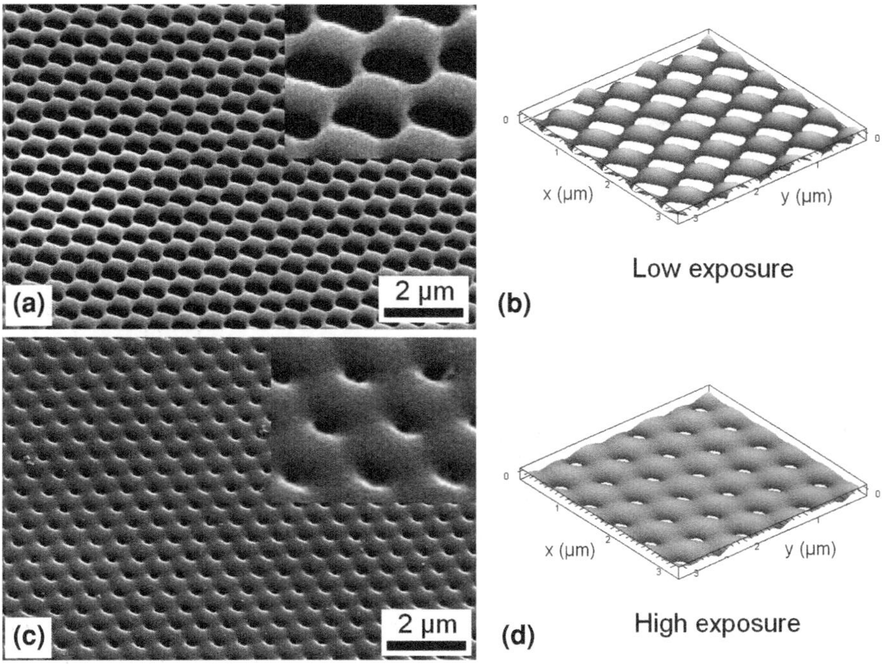

Fig. 7 Periodic two-dimensional hexagonal patterns (rotation angle: 60°) with periodic lattice constant $\Lambda \sim 500$ nm. The exposure doses were (**a**) 3.8 and (**c**) 9.2 mJ/cm^2. The correspondent calculations of solidified material are shown in (**b**) and (**d**) for (**a**) and (**b**), respectively. Reprinted figure with permission from Lasagni et al. [26], Copyright (2010) by WILEY–VCH Verlag GmbH & Co (Editor Jörn Ritterbusch)

In general, in two-beam LIL, the sample is located at the plane $z = 0$ (Fig. 1) and the polar angles of the individual beams are equal or almost identical. Under this condition, vertical polymerized walls perpendicular to the sample surface are obtained. However, if the sample is rotated with a significant rotation angle, the polymerized walls will not be perpendicular to the substrate surface but presenting a specific inclination depending on the incident beam angles (α_i), the rotation angle with respect to the surface (θ), as well as the refraction index of the resist (n_{res}). This setup has been called "non-orthogonal laser interference exposure" method [26].

The real interference angle of the laser beams inside the resist ($\alpha_{i\ \text{res}}$) can be calculated using the Snell's law:

$$\frac{\sin\ \alpha_i}{\sin\ \alpha_{i\ \text{res}}} = \frac{n_i}{n_{\text{res}}} \tag{5}$$

The inclination of the polymerized walls is given by the mean value of both beams inside the resist respect to the substrate. Then, the sample can be rotated but in the opposite direction ($-\theta$) and thus obtaining sub-micrometer canals as shown

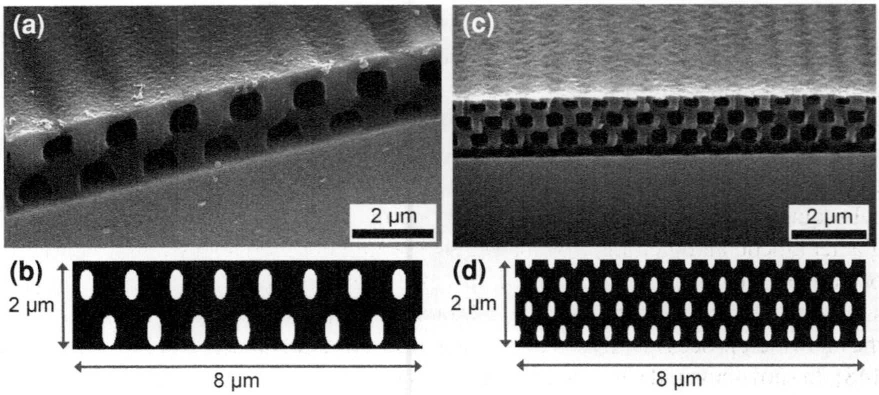

Fig. 8 Fabricated (**a**, **c**) and simulated (**b**, **d**) canal-like structures using the non-orthogonal exposure method. The canal dimensions are **a** width: 1.11; height: 1.11 μm (layer thickness: 2.32 μm) and **c** width: 0.46; height: 0.66 μm (layer thickness: 2.21 μm). The exposure dose in each irradiation step was 8.0 and 5.9 mJ/cm^2 for (**a**) and (**b**), respectively. Reprinted figure with permission from Lasagni et al. [26], Copyright (2010) by WILEY–VCH Verlag GmbH & Co (Editor Jörn Ritterbusch)

in Fig. 8a and c. The obtained structures with sub-micrometer dimensions are difficult to fabricate using conventional photolithographic processes due to different causes such as diffraction effects of the light passing through a mask with small features [43]. These devices can be utilized for applications in microfiltering and micromixing which are normally fabricated using conventional micromachining but with significant larger pore sizes [44].

The achievable mesh size and canal shape are determined by the exposure dose, spatial period and tilted angle. In Fig. 8a, the canals are 1.11 ± 0.05 and 1.12 ± 0.03 μm wide and tall, respectively, while in Fig. 8c the width and height are 0.48 ± 0.02 and 0.66 ± 0.02 μm, in that order [26].

For the case depicted in Fig. 8a, the sample was irradiated with two-beam interference patterns and incident angles α_1 and α_2 of 35 and 55°, respectively (this setup corresponds to the condition $\alpha_1 = \alpha_2 = 10°$, $\beta_1 = 0$, $\beta_2 = \pi$, and rotation angle $\theta = 45°$). Then using Eq. 5 the angles of the beam inside the resist ($n_{res} = 1.6$) are 21.01 and 30.80° and the equivalent rotation angle inside the resist is 25.91° (this angle is the mean value of both incident angles). Finally, the calculated period of the canal-structure parallel to the surface is 1.98 μm that is very similar to the experimental value of 1.81 μm (Fig. 8a). The same analysis can be performed for smaller structure as shown in Fig. 8b. In this case, the calculated period of the canal-structure parallel to the surface is 1.02 μm, which is also very close to the experimental value of 0.91 μm.

To determine the 3D shape of the canals, we can calculate the intensity distribution of the line-like patterns at the incident plane using Eq. 3 with the corrected incident angles (Eq. 5). Then, we need to rewrite the Beer–Lambert's law (Eq. 4) to locally calculate the exposure dose after the beams travel a distance Δl:

$$I(z, x) = I(z - \Delta z, x - \Delta x)\ e^{-\Delta l\ \alpha} \tag{6}$$

$$\Delta l = \Delta x \cos\left(\frac{\alpha_{1\mathrm{res}} + \alpha_{2\mathrm{res}}}{2}\right), \quad \Delta l = \Delta z \cos\left(\frac{\alpha_{1\mathrm{res}} + \alpha_{2\mathrm{res}}}{2}\right) \tag{7}$$

Using this model, the shape of the canals in the irradiated resist was calculated as shown in Fig. 8b and d. In both cases, the expected number of rows was experimentally observed for both conditions: 2 and 4 for Fig. 8b and d, respectively. An important observation for the non-orthogonal laser interference exposure configuration is that it is not possible to determine previous to the experiment the relative position of the interference maxima with respect to the first and second irradiation steps. Nevertheless, for small canal structure features as well as tick resist layers this observation is irrelevant [26].

As previously mentioned, in order to directly frabicate a 3D pattern, at least four coherent beams are necessary ($N - 1$). A particularly interesting application of such 3D patterns is photonic crystals [12]. Multiple interference between waves scattered from each unit cell of the structure may open a "photonic band gap"; a range of frequencies within which no propagating electromagnetic modes exist. Numerous device principles that exploit this property have been identified. Specifically related to interference lithography, 3D polymer structures with submicrometer periodicity have been fabricated as template and subsequently infiltration with high-refractive-index materials. This 3D-periodic structure is in general generated by interference of four non-coplanar laser-beams in a film of photoresist of some tens of micrometers. For the four beam configuration, the intensity distribution of the interference pattern 3D translational symmetry; its primitive reciprocal lattice vectors are equal to the differences between the wave vectors of the beams. Highly exposed photoresist is rendered insoluble. Unexposed areas are dissolved away to reveal a three dimensionally periodic structure formed of cross-linked polymer with air-filled voids. The four laser-beam wave vectors determine the translational symmetry and lattice constant of the interference pattern. There remain eight parameters, describing the intensities and polarisation vectors of the four beams that are required to define the intensity distribution within a unit cell. These parameters allow considerable freedom in determining the distribution of dielectric material within the unit cell, which in turn determines the photonic band structure.

In Fig. 9 different SEM micrographs of photonic crystals with fcc-like symmetry generated by holographic lithography are shown [45]. In this example, Cu_2O (high refractive index material) was electrodeposited into the photopolymerized SU-8 material over a glass substrate covered with a thin ITO layer. Then, the polymer template was removed by isotropic O_2 reactive-ion etching. As shown in Fig. 9a and b, a highly ordered 3D porous structure of Cu_2O was successfully formed. However, although the images reveal that the Cu_2O structure has

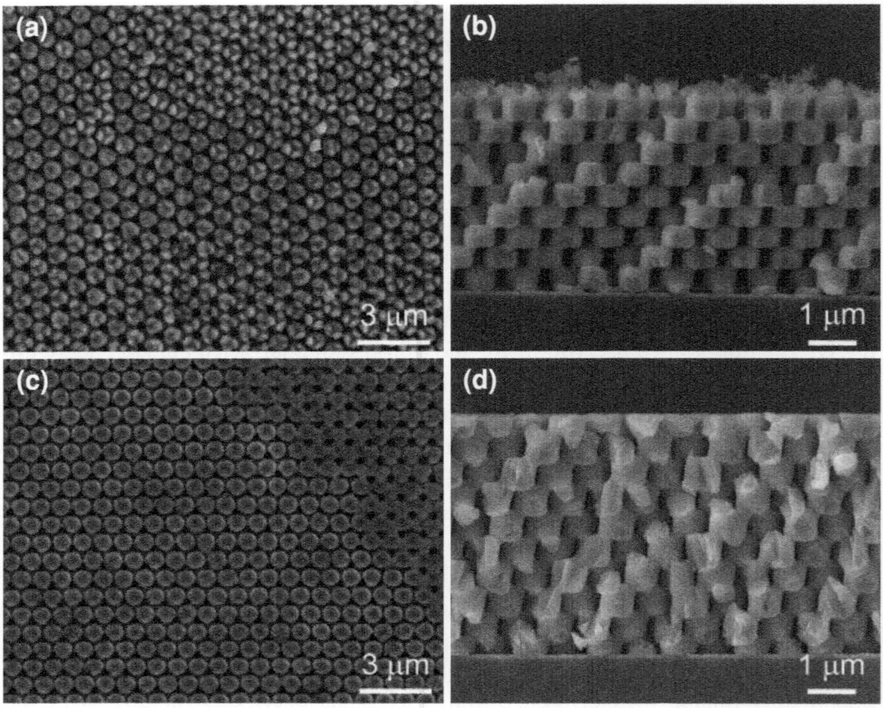

Fig. 9 SEM images of inverse 3D Cu_2O photonic crystals after removal of polymer template. **a** Top surface and **b** fractured cross-section of a Cu_2O photonic crystal electrodeposited within the template followed by Reactive Ion Etching of the template. **c** Top surface and **d** fractured cross section of a polished Cu_2O photonic crystal. Reprinted figure with permission from Miyake et al. [45], Copyright (2009) by WILEY–VCH Verlag GmbH & Co (Editor Peter Gregory)

well-defined smooth surfaces in the bulk of the structure, the top surface is relatively rough. This rough surface is not desirable because it causes diffuse scattering, which reduces the optical strength of the photonic crystal. To solve this problem, the rough surface can be eliminated using simple mechanical polishing with 50 nm alumina abrasive (Fig. 9c and d) [45].

4 Direct Laser Interference Patterning

As has been introduced in Sect. 2, in LIL low laser intensities (up to some tens of mJ/cm^2) are utilized to locally activate a photosensitive material. It means that fabrication of periodic patterns using this approach directly on metals, ceramics or polymers can be not realized in one step and additional treatments are necessary. For example, using chemical or plasma etching, the substrate material can be etched at the regions not covered by the resist. After that, the photoresist is totally removed, generally using a specific solvent, obtaining the final structure on the substrate. However, using laser systems that can achieve higher power densities it is possible to direct fabricate 1 and 2D periodic patterns in one step. In the last case, the method has been called DLIP. In DLIP, nanosecond, picosecond or femtosecond pulsed lasers are used in order to reach high energy densities at the interference maxima positions during the ultrashort laser pulse interaction with the material. In this way, high peak powers ranging from MW to several GW can be obtained during irradiation, permitting to locally melt and/or ablate the irradiated substrates. In the case of metals, this method has been also called LIMET [18, 24]. Following, examples for ns- and fs-DLIP are introducing describing the advantages and disadvantages of the methods.

4.1 DLIP Using Nanosecond Laser Pulses

A good option for the construction of a DLIP system with ns-pulsed laser is to utilize YAG lasers with harmonic generators. In this way, it is possible to select an appropriated wavelength at which the material has high absorption, and thus improving the interaction with the laser light. For example, Nd:YAG lasers (1,064 nm) cover wavelengths ranging from the ultraviolet to the infrared spectrum by frequency doubling (532 nm), tripling (355 nm) and quadrupling (266 nm). Moreover, such systems have an impressive coherent length (in the order of some meters) which is necessary to make use of the so call beam-splitter configuration.

The schematic setup for a beam-splitter interference setup is shown in Fig. 10a. As it can be observed the primary laser beam is split into two or more beams to interfere with each other on the sample surface. By the combination of a polarizer and a half-wave plate it is also possible to adjust the intensity impinging on the sample. The intensity can also be adjusted by selecting beam splitters with a specific reflection. Also a mechanical shutter can be used to control the number of pulses reaching the sample. By a two-lens telescope it is also possible to change the diameter of the laser beam and therefore the energy density received on the sample surface. To reduce influences from the optical elements, high flatness of the splitters and mirrors can be also utilized.

Examples of patterned metallic periodic structures with this configuration are shown in Fig. 11. In this case, DLIP was used to fabricate periodic microstructures with different geometries as well as grating periods on metallic substrates. The good UV absorption ($\lambda = 355$ nm) of the metallic substrates assured a proper absorption of laser radiation, and thus a well process ability even using relatively low energy densities. The grating periods used for the experiments were 1 and 5 μm, and the values of energy density (laser fluence) were varied between 0.4 and 1.4 J/cm^2. It is evident from Fig. 11a and b that patterns with 5 μm grating periods are more pronounced than those with 1 μm. In addition, at low and moderate energy densities (Fig. 11a), the topographic structure is well developed compared to surfaces irradiated with higher intensities (Fig. 11c), regardless of the applied grating period.

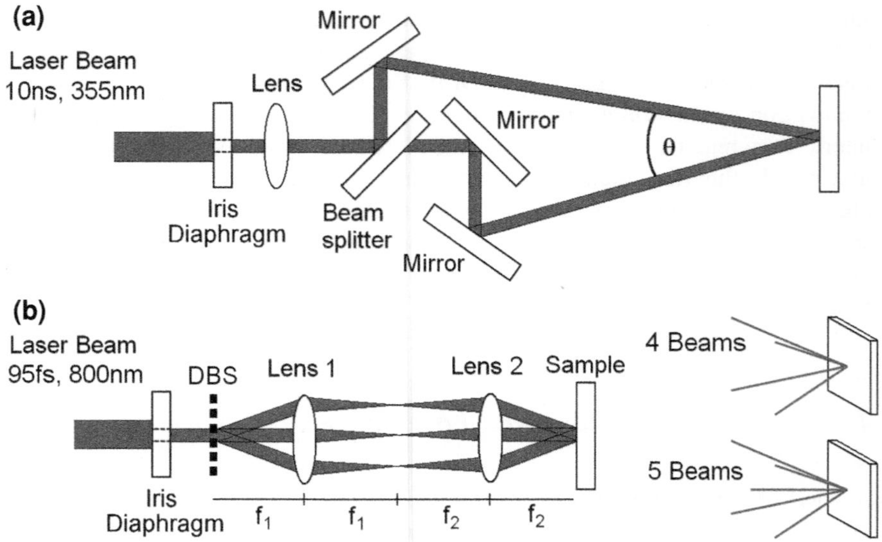

Fig. 10 **a** Experimental setup for nanosecond interference experiments (beam splitter configuration). The principal laser beam is split to obtain two individual laser beams interfering at the sample with an angle θ. **b** Optical setup for the femtosecond interference patterns utilizing a diffractive beam splitter (DBS) and confocal imaging system with two convex lenses with focal distances f_1 and f_2. The four and five laser beam configurations are also shown. Reprinted figure with permission from Lasagni et al. [48], Copyright (2009) by WILEY–VCH Verlag GmbH & Co (Editor Jörn Ritterbusch)

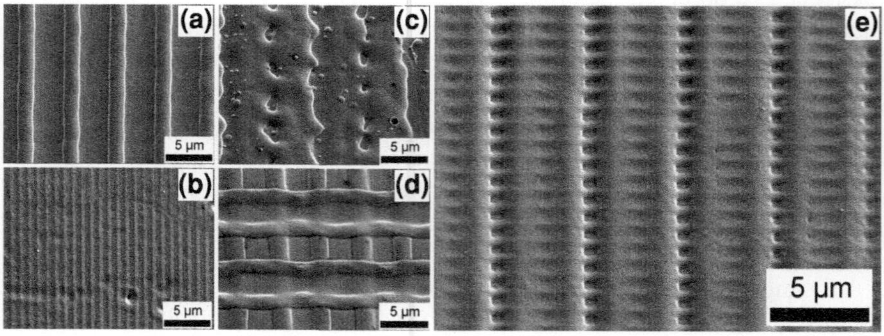

Fig. 11 Periodic line (**a–c**), cross (**d**), and hierarchical patterns (**e**) fabricated on stainless stell and titanium surfaces

The shaping of peak to valley pattern reveals how the melt puddle evolves at the interference maxima (high temperature) and resolidifies at the interference minima (low temperature). Generally, the flow of molten metal is directed away from the intensity maxima towards the intensity minima driven by the surface tension gradient that is induced by the temperature differences between maxima and minima positions [46, 47]. The effect of surface tension-driven convection (also called Marangoni or thermocapillary convection) becomes predominant, if sufficiently high temperature differences between maxima and minima occur.

In addition to line-like patterns, also cross-like structures (Fig. 11d) as well as hierarchical structures (Fig. 11e) can be fabricated by consecutively superimpose interference pattern at different angles (e.g., in Fig. 11e: first grating period: 5 μm, second grating period: 1 μm rotated by 90° with respect to the first grating). The fabrication of such hierarchically patterned surfaces is anticipated to contain several challenges. Specially, the energy density applied for each grating determines the quality of the final structure [49].

A second example is related with the high speed fabrication of organized structures of CNT [50]. Such structures are relevant for the mass production of CNT enabled devices such actuators, field emitters, sensors, cell growth templates, solar cells, as well as LEDs among others [51–56]. While some of these devices require CNT alignment, most require periodic patterning of CNTs on the substrate. Fig. 12 shows CNT films irradiated using interference patterns with spatial periods of (Fig. 12a) 2.86 and (Fig. 12b) 5.73 μm, which were obtained by using angles of 7.1 and 3.6° between the laser beams, respectively. The vertically aligned CNTs were grown using a thermal CVD process in a tube furnace at a temperature of 800°C, and synthesized on a silicon substrate with a 300 nm SiO_2 catalyst support layer.

Keeping the laser fluence (energy density per pulse) fixed at 326 mJ/cm^2, samples of CNTs were exposed to varying numbers of laser pulses. As shown in Fig. 12, regular line-like arrays of CNTs could be fabricated with structure depths up to 29 μm. For this material, the structuring is produced due to the local and periodic ablation of the CNT films at the interference maxima positions. In general, by increasing the number of laser pulses higher structure depth can be

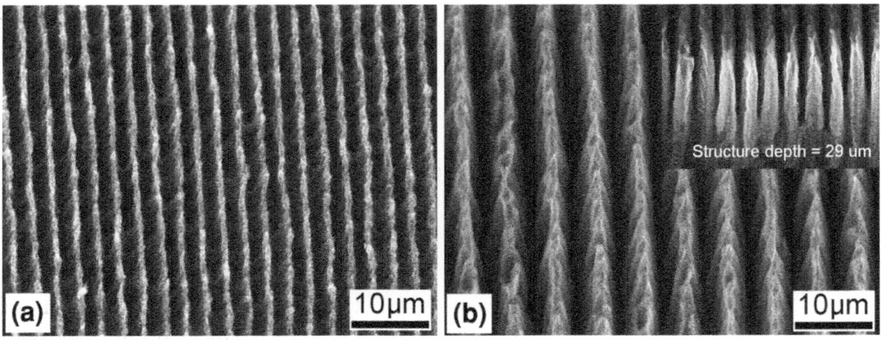

Fig. 12 Scanning electron micrographs of one-dimensional CNT arrays produced by irradiation using interference patterns with spatial periods of (**a**) 2.86 and (**b**) 5.73 μm. In (**a**), only one laser pulse was utilized while 20 pulses were used in (**b**). The inset in (**b**) shows the cross-section of the line-like arrays (structure depth = 29 μm). The laser fluence per pulse was 326 mJ/cm^2. Reprinted figure with permission from Lasagni et al. [50], Copyright (2009) by IOP Publishing Ltd (Editor Mark Reed)

achieved. For example, using one laser pulse with 5.73 μm interference patterns the structure depth was 2.28 μm while for 20 laser pulses the observed value was of 29 μm (see detail in Fig. 12b). In consequence, controlling the number of laser pulses permits to conveniently control the vertical dimension of the desired CNT pattern.

Similarly to LIL, a second approach for the fabrication of 2D structures consisted of performing additional irradiation steps after rotating the sample by a specific angle with respect to the original position. In addition, to improve the quality of the patterns, only one laser pulse per irradiation step was utilized. Fig. 13 shows the morphology of the fabricated 2D patterns. As can be observed, at the positions where the interference maxima are overlapped in the two-step irradiation process, a larger structure depth was observed. The height of the protuberances was around 7.6 μm for both orthogonal and hexagonal patterns (Fig. 13a and c, respectively).

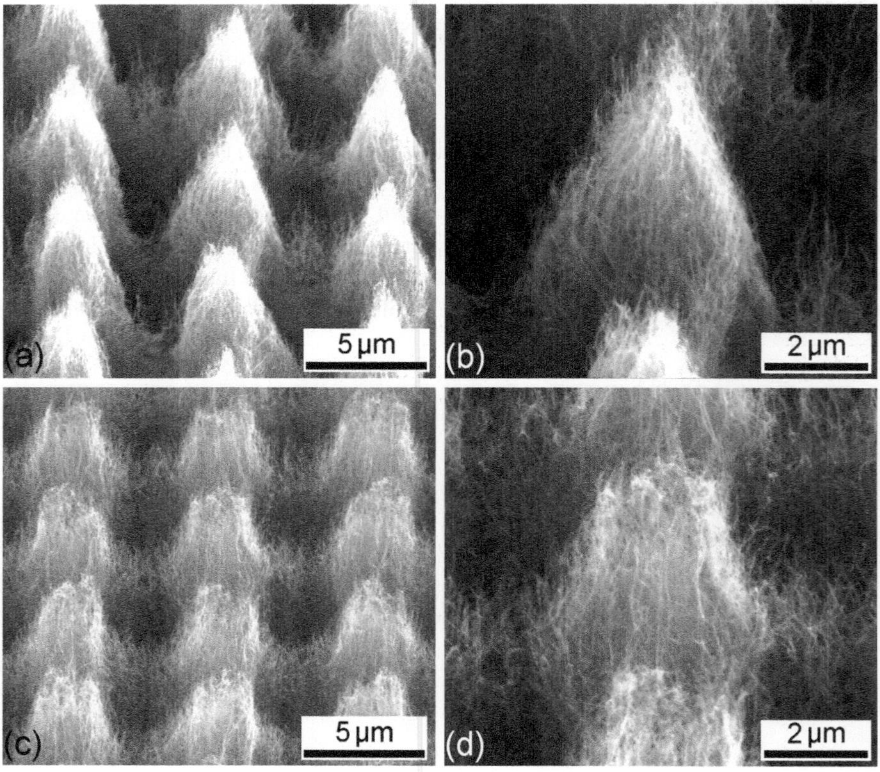

Fig. 13 Scanning electron micrographs of CNT lattice structures fabricated with angles of (**a**, **b**) 60 and (**c**, **d**) 90°. Spatial period: 6.31 μm. The laser fluence per pulse was 305 mJ/cm^2. Reprinted figure with permission from Lasagni et al. [50], Copyright (2009) by IOP Publishing Ltd (Editor Mark Reed)

In addition to controlling the surface topography of the CNT arrays, it is desired that the chemical properties as well as structure of the remained CNT are not changed. Raman spectroscopy can be to properly investigate this effect by observing that the ratio between the graphitic peak (G-peak, around 1,560 cm^{-1}) and the defect peak (D-peak at 1,300 cm^{-1}) [57, 58].Consequently. the greater intensity of the G-peak with respect to the D-peak suggests CNTs with greater purity. For the structures shown in Figs. 12 and 13, this ratio was around 1.16, while for the unpatterned sample this value was 1.15. This clearly suggests that the CNT were removed cleanly and not transformed into amorphous carbon.

4.2 DLIP Using Femtosecond Laser Pulses

It is common knowledge that the shorter the pulses, the smaller the area over which the pulses overlap. For example, in the case of two beams crossed at angle θ, the size of the overlap area (OA) is given by

$$OA = c\tau(\sin\theta/2)^{-1} \tag{8}$$

where c is the speed of light in the medium and τ is the pulse duration [59].

For example, for a 30 fs pulse duration and a moderate angle such as 5°, the beams overlap only within a strip approximately 200 μm wide. The number of interference fringes (N_F) produced by two beams is independent of the angle and, for transform-limited pulses, is roughly:

$$N_F = 2c\tau/\lambda \tag{9}$$

With 30 fs pulses at $\lambda = 800$ nm, only approximately 20 interference fringes can be produced. However, these limitations can be overcome if cross diffraction orders of a grating are crossed using the simple imaging system shown in Fig. 10b [48, 59].

In this setup a femtosecond pulse is transmitted through a diffraction grating, which is then imaged with two confocal lenses, with the grating being placed in the front focal plane of the first lens. A spatial filter transmits only the desired number of laser beams corresponding to the different diffraction orders. Then the beams are recombined at the image plane Fig. 10b. This setup not only provides pulse overlap in the image plane but also preserves short pulse duration [59].

Differently from the beam-splitter configuration (Fig. 10b), in order to change the intercepting angle of the used beams, and in consequence the structure period, a set of lenses with appropriated focal distances (f_1 and f_2) must be selected. Due to the well know interaction of fs lasers with materials, several effects can be explored. One possibility is to fabricate the periodic arrays by direct ablation of the material surface [60–62], and also by photopolymerization processes including multi-photon interactions [48, 63]. This approach has also been utilized to fabricate photonic crystals [64], nano-sized hole matrices and nano-meshes [61] between others.

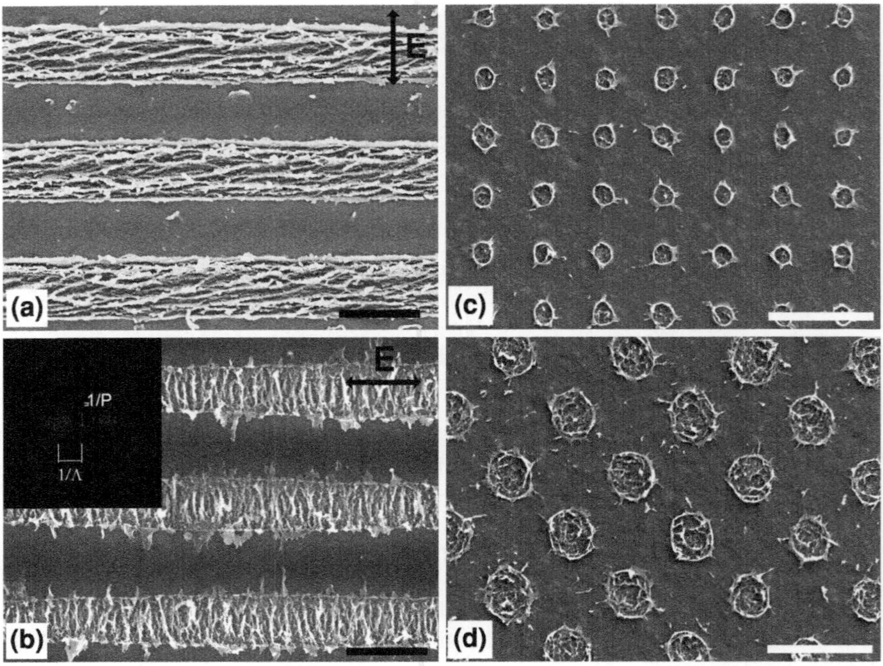

Fig. 14 Fabricated periodic arrays using (**a**, **b**) two, (**c**) four and (**d**) five laser beams configurations with **a** $\alpha_{1,2} = 11°$, $\beta_1 = 0°$, $\beta_2 = 180°$ (two beams); **b** $\alpha_{1,2} = 11°$, $\beta_1 = 90°$, $\beta_2 = 270°$ (two beams); **c** $\alpha_{1,2,3,4} = 11°$, $\beta_1 = 0°$, $\beta_2 = 90°$, $\beta_3 = 180°$, $\beta_4 = 270°$ (four beams); **d** and $\alpha_{1,2,3,4} = 11°$, $\alpha_5 = 0°$, $\beta_1 = 0°$, $\beta_2 = 90°$, $\beta_3 = 180°$, $\beta_4 = 270°$, $\beta_5 = 0°$ (five beams). The accumulated energy density was (**a**, **b**) 0.21 J/cm^2, (**b**) 0.004 J/cm^2, and (**c**) 0.007 J/cm^2. The orientation of the electric field E in (**a**) and (**b**) is denoted by the black. (**a**, **b**) Scale bar = 2 μm; (**c**, **d**) Scale bar = 5 μm. Reprinted figure with permission from Lasagni et al. [21], Copyright (2009) by Elsevier Ltd (Editor F.H.P.M. Habraken)

Examples of patterned periodic structures on poly(3,4-ethylene dioxythiophene)–poly(styrene sulfonate) (PEDOT–PSS) thin films are shown in Fig. 14 [21]. This conducting polymer is one of the most widely used materials in the field of organic electronics [65]. PEDOT–PSS has been used to fabricate neural electrodes [66], flexible photovoltaics [67] as well as batteries [68]. The irradiated films consists on electrochemically deposited PEDOT–PSS conducting polymers from a solution of PSS (0.2%, w/v) and ethylene dioxythiophene (EDOT) (0.1%, w/v) in deionized water, deposited over a 100 nm gold–palladium thin films, sputtered on tin doped indium oxide (ITO) substrates.

To fabricate the arrays, a Ti:Sapphire fs-laser system with a pulse duration of 95 fs, and a center wavelength of 800 nm was utilized. To split the main beam (Fig. 10b) a polycarbonate diffractive beam splitter (DBS) was used. The zero and/or first-order beams were focused onto the sample by two bi-convex lenses with magnification factor $M = f_2/f_1 = 20/75.6 = 0.265$. To avoid distortion of the

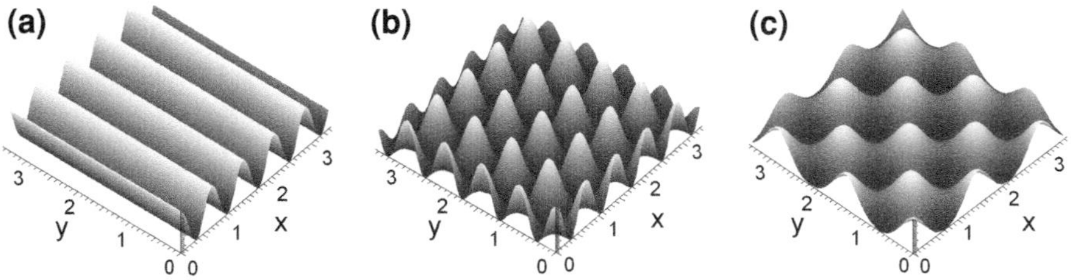

Fig. 15 Calculated intensity distribution by the interference of (**a**) two, (**b**) four, and (**c**) five beams with all the laser beams having the same phase (normalized units). Reprinted figure with permission from Lasagni et al. [21], Copyright (2009) by Elsevier Ltd (Editor F.H.P.M. Habraken)

beam wave fronts due to nonlinear interactions of the high field ultrashort pulses with air, a low humidity atmosphere is required.

Using two, four and five laser beams configurations, different periodic arrays were fabricated (Fig. 14). These fabricated arrays are in agreement with the calculated interference patterns using Eq. 3, as shown in Fig. 15. In addition, the fabricated arrays have been observed on relatively large area (1.28 mm^2) corresponding to the beam diameter at the sample position. The measured spatial period of the periodic arrays is 2.87, 2.89 and 4.08 μm (diagonal) for Fig. 14a–b, c and d, respectively. These values are in good agreement with the calculated periods of 2.96 (two and four beams) and 4.19 μm (five beams) with $\alpha_{1,2} = 11°$, $\beta_1 = 0°$, $\beta_2 = 180°$ (two beams); $\alpha_{1,2} = 11°$, $\beta_1 = 90°$, $\beta_2 = 270°$ (two beams); (c) $\alpha_{1,2,3,4} = 11°$, $\beta_1 = 0°$, $\beta_2 = 90°$, $\beta_3 = 180°$, $\beta_4 = 270°$ (four beams); and $\alpha_{1,2,3,4} = 11°$, $\alpha_5 = 0°$, $\beta_1 = 0°$, $\beta_2 = 90°$, $\beta_3 = 180°$, $\beta_4 = 270°$, $\beta_5 = 0°$ (five beams).

In order to study the effect of laser fluence as well as accumulated energy density over a number of laser pulses on the morphology of the fabricated structures, PEDOT:PSS samples were irradiated with both four and five beam interference patterns. Using relatively high laser fluences (6.7–24.1 mJ/cm^2) we were able to fabricate well defined craters arrays even using only a single laser pulse. The morphology of the craters can be observed in Fig. 14d (five beam configuration, 6.7 mJ/cm^2, 1 laser pulse). In this case, the diameter of the craters was around 2 μm. As the number of laser pulses increases up to ten, several concentric rings within the crater were observed without a significant change in the crater external diameter. The evolution of the crater diameter as function of the accumulated energy density is shown in Fig. 16. As observed, for high energy pulses, the diameter varied from 1.9 to 2.75 μm and 1.8 to 2.4 μm for the four and five beam configurations, respectively. In addition, to completely remove the PEDOT:PSS layer at the interference maxima positions, a substantial number of laser pulses has to be utilized. For example, ~40 and 350 pulses were necessary to completely remove the polymeric layer when using laser fluences of 6.7 and 12.1 mJ/cm^2 for the five beam configuration, respectively. Only when relative high laser fluences per pulse were used (24.1 mJ/cm^2), small numbers of laser pulses

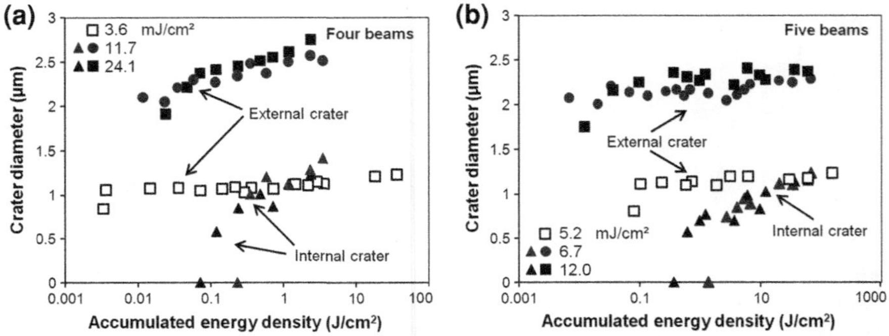

Fig. 16 Crater size evolution as function of accumulated energy density and pulse energy for (**a**) four-beam configuration and (**b**) five-beam configuration. External crater refers to surface modified area while the internal crater corresponds to the total ablated area. Reprinted figure with permission from Lasagni et al. [21], Copyright (2009) by Elsevier Ltd (Editor F.H.P.M. Habraken)

(six laser pulses for the four laser beam configuration) were enough to totally ablated the PEDOT:PSS layer at the interference maxima. In all cases, no damage to the under metallic layer was induced.

Surface damage of PEDOT:PSS is believed to be a combination of linear and nonlinear processes. The debris formation can be explained as the residual preference for etching PEDOT over PSS as has been already suggested by other authors [69]. In addition, the existence of an incubation process can be associated to the ablation process since a certain energy density has to be delivered to start observable damage under sub-threshold conditions when using low energetic pulses (fluences of 3.6 and 5.2 mJ/cm^2) [70].

As shown in Fig. 16a, the smallest size reported of the internal craters for the four beam configuration was 0.58 µm for five laser pulses with individual laser fluence of 24.1 mJ/cm^2 and an accumulated energy density of 120 mJ/cm^2. For the five beam configuration (Fig. 16b), the smallest internal crater size observed was 0.57 µm for 50 pulses with individual laser fluences of 12.0 mJ/cm^2 and an accumulated energy density of 0.61 J/cm^2. When comparing the accumulated energy density necessary for the total ablation of the PEDOT:PSS layer, it can be observed that high energy pulses are more efficient. This effect was observed for both four and five laser beam configurations. A different behavior was observed when irradiating the samples with pulses with a low laser fluence. Even at high accumulated energy densities it was not possible to totally ablate the polymeric layer. In addition, the size of the surface modified areas varied from 1.06 to 1.23 µm for the four beam configuration (pulse laser fluence: 3.6 mJ/cm^2). For the five beam configuration, the crater diameter varied from 1.10 to 1.25 µm. Only when using very few laser pulses, we were able to observe sub-micrometer features (e.g. ~0.76 µm for the five-beam configuration with 17 pulses). Considering that ablation of the conducting polymer occurs when certain threshold laser intensity is surpassed, low energy laser pulses only partially modify the surface topography of the polymeric material at the interference maxima positions without substantial ablation [71, 72].

Another important characteristic of the obtained periodic arrays is the sub-wavelength ripple structure when using several laser pulses (Fig. 14a, b). In all cases, the ripple structure is perpendicular to the beam polarization, and independently of the number of laser beams used in interference set-up as well as accumulated energy density, the mean spatial period of the ripples was between 170 and 220 nm. The orientation of the sub-ripples structure is also denoted by the Fourier transform in Fig. 14b. A particular mechanism that can be regarded to explain the formation of these subwavelength ripples is a Coulombian explosion followed by self-organization [73]. The Coulombian explosion is the result of multiphoton ionization induced by the ultrashort laser pulses. In addition, inter-pulse mechanism could occur in the formation of subwavelength ripples if several pulses are applied [74]. However, as the PEDOT:PSS polymer is conductive, other processes may also considered [74–76].

Acknowledgments The author gratefully acknowledges the contributions of: M. Beida, B.S Menéndez-Ormaza from the Fraunhofer Institute for Material and Beam Technology (Germany); D. Yuan, P. Shao, S. Das, R. Crosss, S. Graham from Georgia Institute of Technology (USA), and J. Hendricks, C.M. Shaw and D. Martin from the University of Michigan (USA). This work was financially supported by Alexander von Humboldt Foundation and the Fraunhofer Association (Grant No. Attract 692174).

References

1. Ganesh, N., Block, I.D., Cunningham, B.T.: Near ultraviolet-wavelength photonic-crystal biosensor with enhanced surface-to-bulk sensitivity ratio. Appl. Phys. Lett. **89**, 023901–023904 (2006)
2. Yu, H., Balogun, O., Li, B., Murray, T.W., Zhang, X.: Building embedded microchannels using a single layered SU-8, and determining Young's modulus using a laser acoustic technique. J. Micromech. Microeng. **14**(11), 1576–1584 (2004)
3. Ryu, W.H., Min, S.W., Hammerick, K.E., Vyakarnam, M., Greco, R.S., Prinz, F.B., Fasching, R.J.: The construction of three dimensional micro-fluidic scaffolds of biodegradable polymers by solvent vapor based bonding of micro-molded layers. Biomaterials **28**, 1174–1184 (2007)
4. Bhatia, S.N., Chen, C.S.: Tissue engineering at the micro-scale. Biomed. Microdev. **2**, 131–141 (1999)
5. Hammerick, K., Ryu, W., Fasching, R., Bai, S., Smith, R., Greco, R.: Synthesis of Cell Structures. In: Greco, R.S., Prinz, F.B., Smith, R.S. (eds.) Nanoscale Technology in Biological Systems, pp. 73–101. CRC Press, Boca Raton (2005)
6. Zhang, J., Venkataramani, S., Xu, H., Song, Y.-K., Song, H.K., Palmore, G., Fallon, J., Nurmikko, A.V.: Combined topographical and chemical micropatterning of neural template for cultured hippocampal neurons. Biomaterials **27**, 5734 (2006)
7. Joannopoulos, J.D., Villeneuve, P.R., Fan, S.: Photonic crystals: putting a new twist on light. Nature **386**, 143–149 (1997)
8. Duarte, M., Lasagni, A., Giovanelli, R., Narciso, J., Louis, E., Mücklich, F.: Increasing lubricant lifetime by grooving periodical patterns using laser interference metallurgy. Adv. Eng. Mater. **10**, 554–558 (2008)

9. Lasagni, A., Nejati, M.R., Clasen, R., Mücklich, F.: Periodical surface structuring of metals by laser interference metallurgy as a new fabrication method of textured solar selective absorbers. Adv. Eng. Mater. **6**, 580–584 (2006)
10. Bertsch, A., Jiguet, S., Renaud, P.: Microfabrication of ceramic components by microstereolithography. J. Micromech. Microeng. **14**, 197–203 (2004)
11. Meriche, F., Neiss-Clauss, E., Kremer, R., Boudrioua, A., Dogheche, E., Fogarassy, E., Mouras, R., Bouabellou, A.: Micro structuring of LiNbO3 by using nanosecond pulsed laser ablation. Appl. Surf. Sci. **254**, 1327–1331 (2007)
12. Campbell, M., Sharp, D.N., Harrison, M.T., Denning, R.G., Turberfield, A.J.: Fabrication of photonic crystals for the visible spectrum by holographic lithography. Nature **404**, 53–56 (2000)
13. Kondo, T., Juodkazis, S., Mizeikis, V., Misawa, H.: Holographic lithography of periodic two- and three-dimensional microstructures in photoresist SU-8. Opt. Exp. **14**, 7943–7953 (2006)
14. Lasagni, A., Holzapfel, C., Weirich, T., Mücklich, F.: Laser interference metallurgy: a new method for periodic surface microstructure design on multilayered metallic thin films. Appl. Surf. Sci. **253**, 8070–8074 (2007)
15. Zhu, X., Xu, Y., Yang, S.: Distortion of 3D SU8 photonic structures fabricated by four-beam holographic lithography with umbrella configuration. Opt. Exp. **15**, 16546–16560 (2007)
16. Miklyaev, Yu, V., Meisel, D.C., Blanco, A., von Freymann, G., Busch, K., Koch, W., Enkrich, C., Deubel, M., Wegener, M.: Three-dimensional face-centered-cubic photonic crystal templates by laser holography: fabrication, optical characterization, and band-structure calculations. Appl. Phys. Lett. **82**, 1284–1286 (2003)
17. Tan, C., Peng, C.S., Petryakov, V.N., Verevkin, Y.K., Zhang, J., Wang, Z., Olaizola, S.M., Berthou, T., Tisserand, S., Pessa, M.: Line defects in two-dimensional four-beam interference patterns. New J. Phys. **10**, 023023 (2008)
18. Mücklich, F., Lasagni, A., Daniel, C.: Laser interference metallurgy-periodic surface patterning and formation of intermetallics. Intermetallics **13**, 437–442 (2005)
19. Lasagni, A., Hendricks, J.L., Shaw, C.M., Yuan, D., Martin, D.C., Das, S.: Direct laser interference patterning of poly(3,4-ethylene dioxythiophene)-poly(styrene sulfonate) (PEDOT-PSS) thin films. Appl. Surf. Sci. **255**, 9186–9192 (2009)
20. Lasagni, A.: Large area fabrication of micro and nano periodic structures on polymers by direct laser interference patterning. In: Proceedings of the 17. Neues Dresdner Vakuumtechnisches Kolloquium, Dresden, October 30–31, 15–19 (2009)
21. Lasagni, A., Shao, P., Hendricks, J.L., Shaw, C.M., Yuan, D., Martin, D.C., Das, S.: Direct fabrication of periodic patterns with hierarchical sub-wavelength structures on poly(3,4-ethylene dioxythiophene)–poly(styrene sulfonate) thin films using femtosecond laser interference patterning. Appl. Surf. Sci. **256**(6), 1708–1713 (2010)
22. Zheng, M., Yu, M., Liu, Y., Skomski, R., Liou, S.H., Sellmyer, D.J., Petryakov, V.N., Verevkin, Yu, K., Polushkin, N.I., Salashchchenko, N.N.: Salashchchenko, Magnetic Nanodot Arrays Produced by Direct Laser Interference Lithography. Appl. Phys. Lett. **79**, 2606 (2001)
23. Lasagni, A., Mücklich, F.: Study of the multilayer metallic films topography modified by laser interference irradiation. Appl. Surf. Sci. **240**, 214–221 (2005)
24. Mücklich, F., Lasagni, A., Daniel, C.: Laser interference metallurgy - using interference as a tool for micro/nano structuring. Zeit. für Metallk. **97**, 1337–1344 (2006)
25. Mützel, M., Tandler, S., Haubrich, D., Meschede, D., Peithmann, K., Flaspöhler, M., Buse, K.: Atom lithography with a holographic light mask. Phys. Rev. Lett. **88**(8), 83601 (2002)
26. Lasagni, A., Menéndez-Ormaza, B.: How to fabricate two and three dimensional micro and sub-micrometer periodic structures using two-beam Laser Interference Lithography. Adv. Eng. Mater. **12**(1–2), 54–60 (2010)
27. Cheng Lu, R., Lipson, H.: Interference lithography: a powerful tool for fabricating periodic structures. Laser Photon. Rev. **4**, 568–580 (2009)
28. Moon, J.H., Ford, J., Yang, S.: Fabricating three-dimensional polymer photonic structures by multi-beam interference lithography. Polym. Adv. Technol. **17**(2), 83–93 (2006)

29. Seidemann, V., Rabe, J., Feldmann, M., Buttgenbach, S.: SU8-micromechanical structures with in situ fabricated movable parts. Microsys. Technol. **8**(4), 348–350 (2002)
30. Rubner, R.: Photoreactive polymers for electronics. Adv. Mater. **2**(10), 452–457 (1990)
31. Kuiper, S., van Wolferen, H., van Rijn, C., Nijdam, W., Krijnen, G., Elwenspoek M.: Fabrication of microsieves with sub-micron pore size by laser interference lithography. J. Micromech. Microeng. **11**(1), 33–37 (2001)
32. Carter, J.M., Fleming, R.C., Savas, T.A., Walsh, M.E., O'Reilly, T.B., Schattenburg, M.L., Smith, H.I.: Interference lithography. MTL Annual report (2003)
33. Hoffnagle, J.A., Hinsberg, W.D., Sanchez, M., Houle, F.A.: Liquid immersion deep-ultraviolet interferometric lithography. J. Vac. Sci. Technol. **B17**, 3306–3309 (1999)
34. Switkes, M., Rothschild, M.: Immersion lithography at 157nm. J. Vac. Sci. Technol. **B19**, 2353–2356 (2001)
35. Burnett, J.H., Kaplan, S.G.: Measurement of the refractive index and thermooptic coefficient of water near 193 nm. J. Microlith. Microfab. Microsys. **3**, 68–72 (2004)
36. Raub, A.K., Brueck, S.R.J.: Deep UV immersion interferometric lithography. Proc. SPIE **5040**, 667–678 (2003)
37. Raub, A.K., Frauenglass, A., Brueck, S.R.J., Conley, W., Dammel, R., Romano, A., Sato, M., Hinsberg, W.: Imaging capabilities of resist in deep-UV liquid immersion interferometric lithography. J. Vac. Sci. Technol. **22**, 3459–3464 (2004)
38. Divliansky, I.B., Shishido, A., Khoo, I., Mayer, T.S., Pena, D., Nishimura, S., Keating, C.D., Mallouk, T.E.: Fabrication of two dimensional photonic crystals using interference lithography and electrodeposition of CdSe. Appl. Phys. Lett. **79**, 3392–3394 (2001)
39. Lu, C., Hu, X., Mitchell, I., Lipson, R.: Diffraction element assisted lithography: Pattern control for photonic crystal fabrication. Appl. Phys. Lett. **86**, 193111–193113 (2005)
40. Wu, D., Fang, N., Sun, C., Zhang, X.: Stiction problems in releasing of 3D microstructures and its solution. Sens. Actuators A **128**, 109–115 (2006)
41. Vora, K.D., Peele, A.G., Shew, B.Y., Harvey, E.C., Hayes, J.P.: Fabrication of support structures to prevent SU-8 stiction in high aspect ratio structures. Microsyst. Technol. **13**, 487–493 (2007)
42. Itani, T., Yoshino, H., Hashimoto, S., Yamana, M., Samoto, N., Kasama, K.: A study of acid diffusion in chemically amplified deep ultraviolet resist. J. Vac. Sci. Technol. B **14**, 4226–4228 (1996)
43. Yoon, Y.-K., Park, J.-H., Allen, M.G.: Multidirectional UV Lithography for Complex 3-D MEMS Structures. J. Micromech. Microeng. **15**, 1121–1130 (2006)
44. Sato, H., Kakinuma, T., Go, J.S., Shoji, S.: In-channel 3-D micromesh structures using maskless multi-angle exposure and their microfilter application. Sens. Actuators A **111**, 87–92 (2004)
45. Miyake, M., Chen, Y.-C., Braun, P.V., Wiltzius, P.: Fabrication of three-dimensional photonic crystals using multibeam interference lithography and electrodeposition. Adv. Mater. **21**, 3012–3015 (2009)
46. Drezet, J.M., Pellerin, S., Bezençon, C., Mokadem, S.: Modelling the Marangoni convection in laser heat treatment. J. Phys. IV France **120**, 299–306 (2004)
47. von Allmen, M., Blatter, A.: Laser-beam interactions with materials–physical principles and applications, 2nd edn. Springer-verlag, Heidelberg (1995)
48. Lasagni, A., Yuan, D., Shao, P., Das, S.: Periodic Micropatterning of polyethylene glycol diacrylate hydrogel by laser interference lithography using nano- and femtosecond pulsed lasers. Adv. Eng. Mater. **11**(3), B20–B24 (2009)
49. Lasagni, A., Cross, R., Graham, S., Das, S.: Fabrication of high aspect ratio carbon nanotube arrays by direct laser interference patterning. Nanotechnology **20**, 245305–245312 (2009)
50. Bonard, J.M., Salvetat, J.P., Stockli, T., de Heer, W.A., Forro, L., Chatelain, A.: Field emission from single-wall carbon nanotube films. Appl. Phys. Lett. **73**, 918–920 (1998)
51. Choi, W.B., Chung, D.S., Kang, J.H., Kim, H.Y., Jin, Y.W., Han, I.T., Lee, Y.H., Jung, J.E., Lee, N.S., Park, G.S., Kim, J.M.: Fully sealed, high-brightness carbon-nanotube field-emission display. Appl. Phys. Lett. **75**, 3129–3131 (1999)

52. Dai, H., Franklin, N., Han, J.: Exploiting the properties of carbon nanotubes for nanolithography. Appl. Phys. Lett. **73**, 1508–1510 (1998)
53. Fan, S., Chapline, M.G., Franklin, N.R., Tombler, T.W., Cassell, A.M., Dai, H.: Self-oriented regular arrays of carbon nanotubes and their field emission properties. Science **283**, 512–514 (1999)
54. Wang, Q.H., Setlur, A.A., Lauerhaas, J.M., Dai, J.Y., Seelig, E.W., Chang, R.P.H.: A nanotube-based field-emission flat panel display. Appl. Phys. Lett. **72**, 2912–2913 (1998)
55. Wong, S.S., Joselevich, E., Woolley, A.T., Cheung, C.L., Lieber, C.M.: Covalently functionalized nanotubes as nanometre-sized probes in chemistry and biology. Nature **394**, 52–55 (1998)
56. Bieda, M., Beyer, E., Lasagni, A.F.: Direct fabrication of hierarchical microstructures on metals by means of direct laser interference patterning. J. Eng. Mater. Technol. **132**, 031015–031021 (2010)
57. Jiang, C., Ko, H., Tsukruk, V.V.: Strain sensitive Raman modes of carbon nanotubes in deflecting freely suspended nanomembranes. Adv. Mater. **17**, 2127–2131 (2005)
58. Jorio, A., Pimenta, M.A., Souza Filho, A.G., Saito, R., Dresselhaus, G., Dresselhaus, M.S.: Characterizing carbon nanotube samples with resonance Raman scattering. New J. Phys. **5**, 139–156 (2003)
59. Maznev, A.A., Crimmins, T.F., Nelson, K.A.: How to make femtosecond pulses overlap. Opt. Lett. **23**, 1378–1380 (1998)
60. Nakata, Y., Okada, T., Maeda, M.: Fabrication of dot matrix, comb, and nanowire structures using laser ablation by interfered femtosecond laser beams. Appl. Phys. Lett. **81**, 4239–4241 (2002)
61. Nakata, Y., Okada, T., Maeda, M.: Lithographical laser ablation using femtosecond laser. Appl. Phys. A **79**, 1481–1483 (2004)
62. Tan, B., Sivakumar, N.R., Venkatakrishnan, K.: Direct grating writing using femtosecond laser interference fringes formed at the focal point. J. Opt. A Pure Appl. Opt. **7**, 169–174 (2005)
63. Kondo, T., Matsuo, S., Juodkazis, S., Mizeikis, V.: Multiphoton fabrication of periodic structures by multibeam interference of femtosecond pulses. Appl. Phys. Lett. **82**, 2758–2760 (2003)
64. Kondo, T., Matsuo, S., Juodkazis, S., Misawa, H.: Femtosecond laser interference technique with diffractive beam splitter for fabrication of three-dimensional photonic crystals. Appl. Phys. Lett. **79**, 725–727 (2001)
65. Beaupre, S., Leclerc, M.: Optical and electrical properties of π-conjugated polymers based on electron-rich 3,6-dimethoxy-9,9-dihexylfluorene unit. Macromolecules **36**, 8986–8991 (2003)
66. Yang, J.Y., Kim, D.H., Hendricks, J.L., Leach, M., Northey, R., Martin, D.C.: Ordered surfactant-templated poly(3,4-ethylenedioxythiophene) (PEDOT) conducting polymer on microfabricated neural probes. Acta Biomater. **1**, 125–136 (2005)
67. Aernouts, T., Vanlaeke, P., Geens, W., Poortmans, J., Heremans, P., Borghs, S., Mertens, R., Andriessen, R., Leenders, L.: Printable anodes for flexible organic solar cell modules. Thin Solid Films **22**, 451–452 (2004)
68. Khomenko, V.G., Barsukov, V.Z., Katashinkii, A.S.: The catalytic activity of conducting polymers toward oxygen reduction. Electrochim. Acta **50**, 1675–1683 (2005)
69. Xiao, Y., Takashi, I., Higgins, D.A.: Grayscale patterning of polymer thin films with nanometer precision by direct-write multiphoton photolithography. Langmuir **24**, 8939–8943 (2008)
70. García-Navarro, A., Agulló-López, F., Olivares, J., Lamela, J., Jaque, F.: Femtosecond laser and swift-ion damage in lithium niobate: a comparative analysis. J. Appl. Phys. **103**, 093540 (2008)
71. Sohn, I.-B., Noh, Y.-C., Choi, S.-C., Ko, D.-K., Lee, J., Choi, Y.-J.: Femtosecond laser ablation of polypropylene for breathable film. Appl. Surf. Sci. **254**, 4919–4924 (2008)

72. Vorobyeb, A., Guo, C.: Change in absorptance of metals following multi-pulse femtosecond laser ablation. J. Phys. **59**, 579–584 (2007)
73. Reif, J., Costache, F., Henyk, M., Pandelov, S.V.: Ripples revisited: non-classical morphology at the bottom of femtosecond laser ablation craters in transparent dielectrics. Appl. Surf. Sci. **197–198**, 891–895 (2002)
74. Guillermin, M., Garrelie, F., Sanner, N., Audouard, E., Soder, H.: Single- and multi-pulse formation of surface structures under static femtosecond irradiation. Appl. Surf. Sci. **253**, 8075–8079 (2007)
75. van Driel, H.M., Sipe, J.E., Young, J.F.: Laser-Induced Periodic Surface Structure on Solids: A Universal Phenomenon. Phys. Rev. Lett. **49**, 1955–1958 (1982)
76. Young, J.F., Preston, J.S., van Driel, H.M., Sipe, J.E.: Laser-induced periodic surface structure. II. Experiments on Ge, Si, Al, and brass. Phys. Rev. B **27**, 1155–1172 (1983)

Laser Micromachining

Udo Klotzbach, Andrés Fabián Lasagni, Michael Panzner and Volker Franke

Abstract Over the recent years, the laser technology and its potentials have been exciting laser manufacturers as well as researchers and industrial users. Lasers with their excellent beam quality promised noticeable advantages and improvements in high precision and material processing at the microscale. Besides the excellent beam quality there are more advantages of the laser technology such as compact installation size, high laser efficiency, moderate system price and easy to be integrated. This chapter describes the results of different short pulse laser systems in a wide field of applications, showing specific advantages in practical applications. Typical applications of laser micro machining have been chosen including drilling, cutting, structuring, lateral material removal as well as marking.

1 Introduction

The excellence attributes of laser radiation combined with a high degree off flexibility, contact- and wear-less machining, the possibility of high automation as well as easy integration allows using this tool in a wide field of macro machining processes for many materials including silicon, ceramics, metal and polymer.

The world of laser machining production is divided into micro- and macro-machining. This classification is not based on the size of the work piece but rather the fineness of the impact cased by the laser tool. The lasers system used for micromachining employ normally pulsed beams with an average power of well

U. Klotzbach (✉) · A. F. Lasagni · M. Panzner · V. Franke
Fraunhofer Institute for Material and Beam Technology (IWS),
Winterbergstr. 28, 01277 Dresden, Germany
e-mail: udo.klotzbach@iws.fraunhofer.de

F. A. Lasagni and A. F. Lasagni (eds.), *Fabrication and Characterization in the Micro-Nano Range,* Advanced Structured Materials, 10,
DOI: 10.1007/978-3-642-17782-8_2, © Springer-Verlag Berlin Heidelberg 2011

below 1 kW while those used for macro machining use generally continuous-wave (CW) laser beams ranging up to several kW.

Laser micromachining techniques are currently used by the automobile and medical industries as well as in the production of semiconductors and solar cell processing.

Lasers for micromachining offer a wide range of wavelengths, pulse duration (from femtosecond to microsecond) and repetition rates (from single pulse to Megahertz). These attributes allow micromachining with high resolution in depth and lateral dimensions.

The field of micro-machining includes manufacturing methods like drilling, cutting, welding as well as ablation and material surface texturing, whereby it is possible to achieve very fine surface structures ranging in the micrometer domain. Such processes require a rapid heating, melting and evaporation of the material. The use of extremely short nano- and pico- and even femtosecond pulse durations helps to minimize the thermal effects such as melting and burr formation thus eliminating the need for any post processing measures.

Besides choosing the proper laser source, it is often necessary to employ specialized micromachining components in order to achieve a desired geometry. Currently, laser micromachining is accomplished with two technologies: (i) direct laser writing (DLW) using solid state lasers with a 2D galvanometer scan head, and (ii) mask projection technique (MP) using mostly Excimer lasers and conventional fixed masks.

In DLW, hard or low-sensitive materials generally require solid state lasers in combination with galvanometer scan heads and focussing optics to minimize the beam diameter and guide it over the material surface (Fig. 1).

Based on today's approved scan head technologies and software, even processing of three dimensional surfaces is possible. Advantages are simple programmability and the resulting flexibility. Generation of characters, barcodes and other patterns is easy. Compatibility to existing file formats like "DXF" (Drawing Interchange Format) or "PLT" (Plot File) is given by the software's import features. In consequence, this process can be easily controlled, permitting a high flexibility.

Limitations of the direct writing process are given by the inherent characteristic of the sequential information transfer and dynamics of the scan head. In consequence there is speed limitation that must be considered. A second and often even more important limitation is given by the generally high intensities, which result from high power and tight focusing of the laser beam. Therefore marking of sensitive materials like paper or plastics is hard or even impossible.

For sensitive materials (especially organic materials) often Excimer lasers are used. Compared to frequency tripled solid state lasers their benefit is that even shorter wavelength can be utilized which allows to permanently change the color of UV-absorbing pigments like titanium oxide. These lasers are commonly used in mask projection techniques (MP). Like in photolithography, the mask projection technique enables to transfer all information contained in a fixed transmissive mask at once (Fig. 2). In addition to that, the laser beam must be collimated and

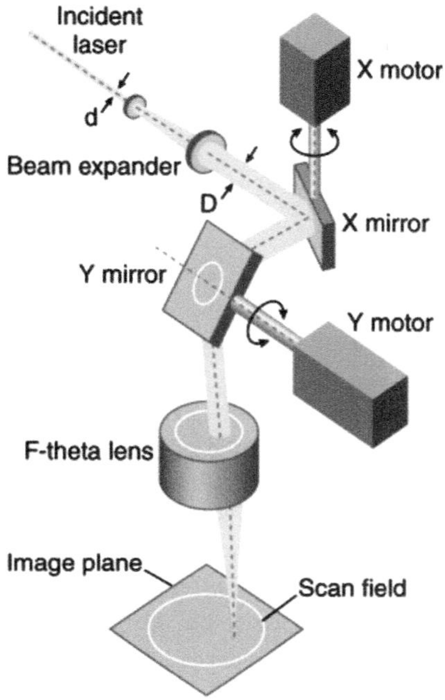

Fig. 1 Principle of direct laser writing method with *xy*-galvanometer scan-head

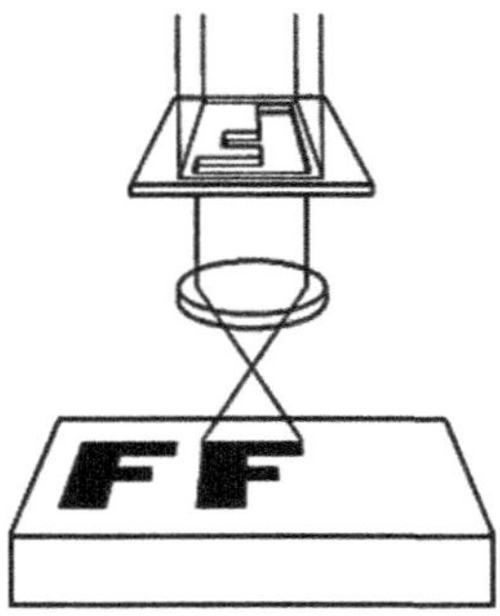

Fig. 2 Principle of mask projection technique

homogenized before illuminating the mask, and the projection system permit to control the size of the projected two dimensional objects.

With this technique a complete marking of alphanumeric characters or pictures is generated within one laser pulse which typically has a length of some nanoseconds. In this way, the process speed is just limited by the repetition rate of the Excimer laser and not by mechanical movement of mirrors with galvanometer scanners. However there is evidently a drawback of the technique, since each new marking requires a new mask.

2 Laser Micro Structuring of Silicon

Micro structures in silicon are applied in different industrial fields as well as in medicine and research. Examples are micro mechanical sensors for car security systems, nozzle plates for printer, and optical elements for X-ray beam splitting. Wherever the accuracy of etched silicon structures is not required, laser processes with short pulses and small wave length can be an option with the advantage of shorter process time.

High quality cutting and drilling of silicon can be achieved by application of diode pumped q-switched Nd:YAG-lasers and harmonics generation. The short pulses (<15 ns) and the UV wavelength (e.g. 266 and 355 nm) for example, allow a reduction of thermal effects like deposition of molten material and heat effected zones (HAZ) at the edges. Especially, in the case of deep structures, the ablation plasma causes strong heating of the walls. A reduction of the plasma temperatures and so the heat influence on the walls can be realized by using smaller laser wavelengths with low plasma absorption. Short laser pulse durations are necessary to reduce the heat affected depth or melting due to heat flow from the ablation area into the bulk material. Also the duration and intensity of plasma heating is reduced by short laser pulses.

In this section the possibilities and limits of laser machining of silicon by diode pumped Nd:YAG-lasers with harmonics generation are discussed.

2.1 Interaction of Laser Radiation with Silicon

Silicon is one of the most thoroughly investigated materials. The vast majority of electronic components are based on semi-conducting mono-crystalline silicon. Polycrystalline and amorphous silicon is used in solar cell technology. Micro-mechanical components as acceleration sensors in car safety systems and micro-fluidic circuits are made of silicon. Silicon is also used as a substrate for X-ray mirrors.

Each of the above mentioned applications involves the structuring of the bulk material since this material cannot be directly processed as pre-formed parts, with the necessary quality requirements. In general, the high accuracy necessary for micro-mechanical components with their structural dimensions of typically a few micrometers is realized with etching technology. Mechanical cutting with its lower accuracy is applied for separating the chips on the wafer. Also laser induced etching [1] is coming up as new emerging process technology. Laser processing has an accuracy intermediate between etching and mechanical cutting, combined with the essential advantage of high flexibility: the beam spot on the material can be rapidly moved by scanning, and/or the sample can be moved below the spot.

The technology applied for a special purpose is chosen according to several quality criteria including:

(i) amount of deposited material along the edges,
(ii) width of damaged material along the edges,
(iii) minimum structural dimensions,
(iv) maximum flank slope, smoothness of flanks and bottom.

The accuracy of structuring with pulse laser depends on wavelength, pulse time, pulse repetition rate, and intensity distribution on the beam spot, as well as on the beam guiding technique.

Much effort has been done in application of short pulse laser technology for silicon processing [2, 3]. When comparing femtosecond laser technology and ns-laser technology, it appears that industrial requirements can be more easily satisfied with ns-lasers.

Improved accuracy of silicon structuring can be expected from using harmonics with their shorter wavelength instead of the original IR-beam of the Nd:YAG-laser. The penetration depth of the beam into the material is usually quantified by a decay length, here denoted as x_{opt}, which is the inverse of the absorption coefficient $\alpha(\lambda)$. Note that its dependence on the wavelength λ may be very strong as seen in Fig. 3. In case of very small penetration depths, there is a high concentration of deposited power per pulse so that essentially the whole heated layer can be removed while the amount of melt can be negligibly small. In case of larger penetration depths and correspondingly lower concentration of deposited power per pulse, a larger amount of melt is present but the rate of removal may be low.

The decay lengths (or optical penetration depth) of the Nd:YAG-laser radiation and its harmonics as derived from Fig. 3 are 60, 0.5 and 0.01 μm for wavelengths of 1064, 532, and 355 nm, respectively.

Whether the optical penetration depth is a relevant quantity here depends on the comparison with the thermal penetration depth x_{th} related to the thermal diffusivity κ of the material

$$x_{th} = (\kappa\tau)^{1/2}, \quad \kappa = \frac{\lambda}{\rho c} \qquad (1)$$

Fig. 3 Absorption coefficient in dependence on wave length [6]

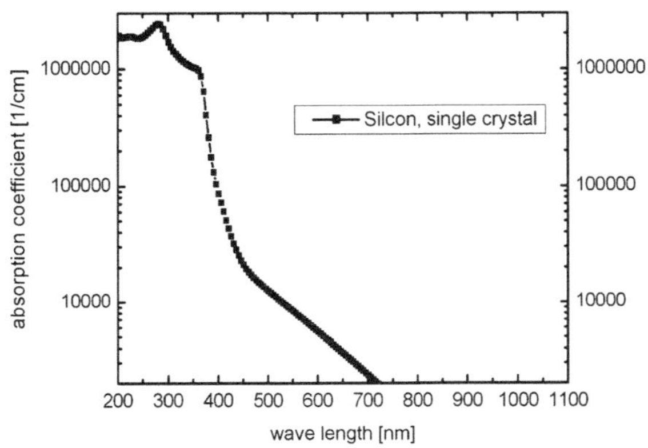

where ρ is the density, c is the specific heat, and τ is the pulse duration

Obviously, shorter pulses lead to smaller thermal penetration depths. For 15 ns pulses, the thermal penetration depth is ~4 μm in silicon.

Apart from that, the accuracy of the generated structure and the quality of its edges are influenced by the ablation plasma. Especially within deep holes, plasma can lead to melt formation at the wall and to removal of material in places where it is not to be removed [4].

To avoid exceedingly high plasma temperature, the absorption by inverse Bremsstrahlung should be kept low. As the plasma absorption α_p goes with the inverse square of the frequency [5],

$$\alpha_p \approx \frac{n}{\nu^2} \cdot e^{(\Delta E_0)/kT_e} \tag{2}$$

where n is the electron density, k is the Boltzmann constant, T_e is the electron temperature, ΔE_0 is the energy gap, and ν is the light frequency.

Higher frequencies or shorter wavelengths produce a lower absorption and hence lower temperatures.

Melt formation may also cause problems as the melt driven out of the hole by the plasma pressure can get stuck to the cooler parts of the borehole wall. From the above considerations it can be deduced that diode-pumped solid-state lasers with harmonics generation and pulses $\tau < 30$ ns, which have become commercially available only recently, should be suitable for high-accuracy structuring processes on silicon.

Next, several examples of micro-processing possibilities using laser radiation on silicon are described. The experiments have been done with a frequency-doubled Nd:YAG-laser and a frequency-tripled Nd:YAG-laser with average powers of 13.5 and 3 W, respectively and a repetition rate of 10 kHz for both cases.

2.2 Single Pulses and Percussion Holes

The focused laser beam spot was moved over the silicon surface by means of scanner with a path velocity as high as 6.5 m/s so that with 10 kHz repetition rate the effect of the individual pulses was resolved, as seen in the SEM images (Fig. 4). A 15 μm wide crater was formed, surrounded by a 40 μm wide area covered with splashed melt.

With stationary spot, through holes are formed after 20,000 pulses (Fig. 5). In this case, the beam was focused on the surface of the sample, producing a hole with 25 μm in diameter at the entry and 20 μm at the exit. Then, the taper ratio (T) for this case was ~0.9% calculated using Eq. 3. The last parameter can be related with the quality of the produced hole:

$$T\% = \frac{d_{\text{entry}} - d_{\text{exit}}}{\text{th}} \cdot 100\% \tag{3}$$

where th is the substrate thickness

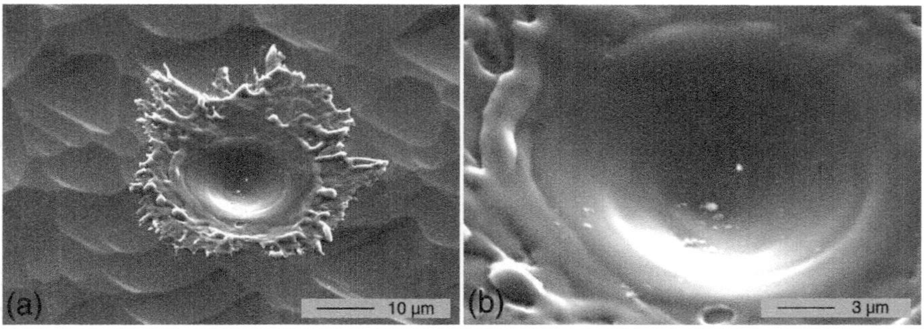

Fig. 4 SEM-images of the effect of single UV pulses on an unpolished Si (111). **a** Surface overview of the whole spot and **b** its internal region with a higher magnification (sample is slightly tilted for better visualisation)

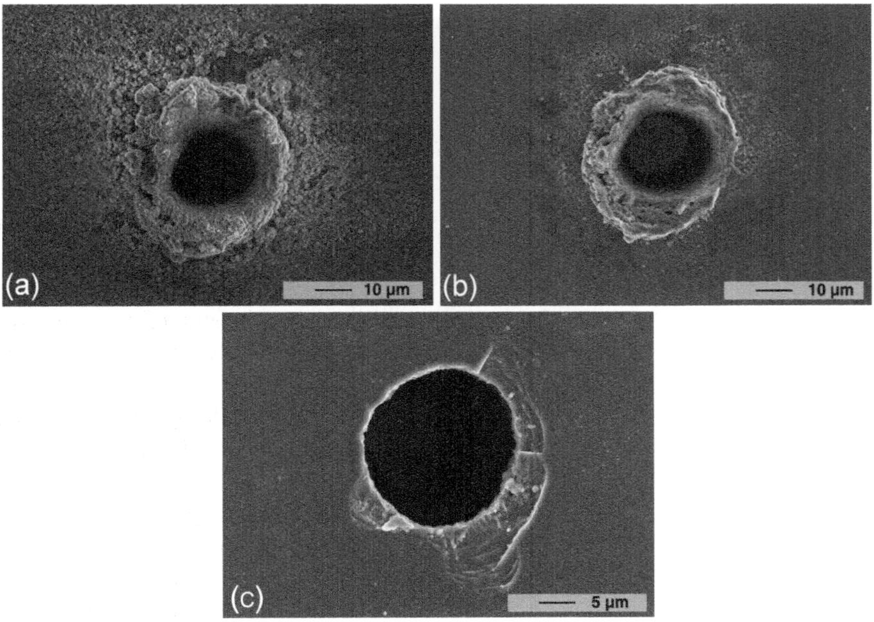

Fig. 5 SEM images of percussion holes in Si (111), thickness 550 µm, 20,000 pulses, 10 kHz: **a** hole entry without HF-cleaning, **b** hole entry after HF-cleaning, **c** hole exit

After structuring, the substrates were etched with 20% hydrofluoric acid and rinsed with distilled water to remove deposits of Silicon oxide along the edges. The effect of hydrogen fluoride (HF) cleaning is distinctly visible: the deposited Silicon oxide is removed. However, the cleaned surface seems to be slightly rough. A molten rim may be also formed around the entry edge of the hole.

2.3 Silicon Cutting

With the short wavelengths and short pulses applied here, the laser beam erosion of Silicon is dominated by sublimation. However, it is obvious that transient melting is also part of the process.

The depth of trenches obtained with repeated erosion cycles is plotted in Figs. 6 and 7 for two beam-path velocities for both wavelengths. The erosion rate has always been found larger at the beginning and becoming nearly stationary, with some fluctuations, at a depth of about 200 μm. In the case of cutting thick samples we found erosion stopping because of enhanced beam reflections at the trench walls for smaller spots. Wider trenches allow deeper cuts.

The cause for this trend is not obvious. As expected from Fig. 3, the erosion rate at 355 nm should be higher than at 532 nm. Because of the low accuracy of depth measurement, the difference in the erosion rate for 5 and 100 mm/s path velocities seems to be insignificant. The cross-shaped cut in a 550 μm silicon wafer (see Fig. 8a) gives an impression of the cut quality, which is even higher for thinner substrates since less heat accumulation and melting near the edges are produced during the cutting process (Fig. 8b). Cutting with 335 nm beam produces smaller amounts of melt and hence higher quality. The obtained gap widths are 40 and 100 μm, respectively.

Figure 8b shows a section through a 550 μm wafer cut with UV-laser (355 nm). The taper angle of the gap is 88.3° and the width is 18 μm below (Fig. 9a), with higher edge quality than above.

In addition to straight cuts or simple geometries, by combining the correct solid-state laser with a flexible and well controllable scanner, structures suitable for micro-mechanics and micro-fluidics can be also easily produced. For demonstration, a toothed gear has been cut from a 550 μm silicon slab (Fig. 9b). The clean edges have been obtained with hydrofluoric acid etching as described above.

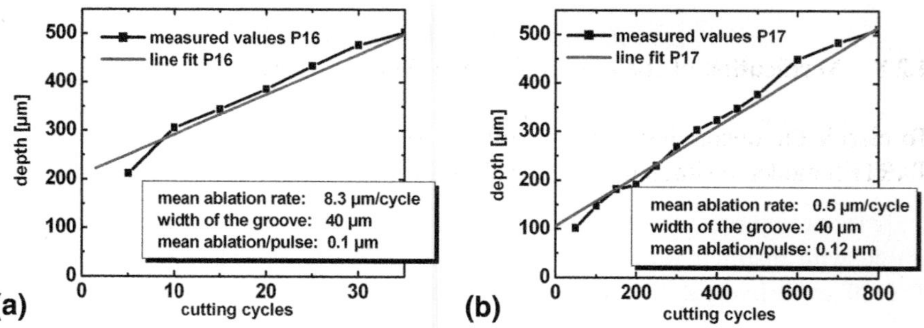

Fig. 6 Cutting depth in dependence on the number of cutting cycles. Sample Si (111), thickness ~550 μm: **a** UV-laser, cutting velocity 5 mm/s; **b** UV-laser, cutting velocity 100 mm/s

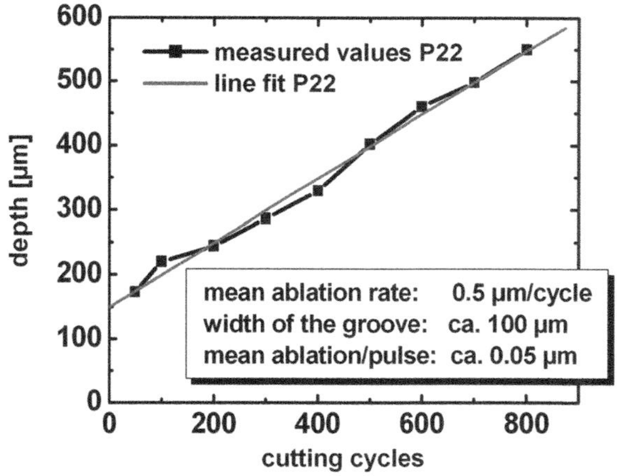

Fig. 7 Cutting depth in dependence on the number of cutting cycles. Sample Si (111), thickness ~550 µm. 532 nm laser, cutting velocity 100 mm/s

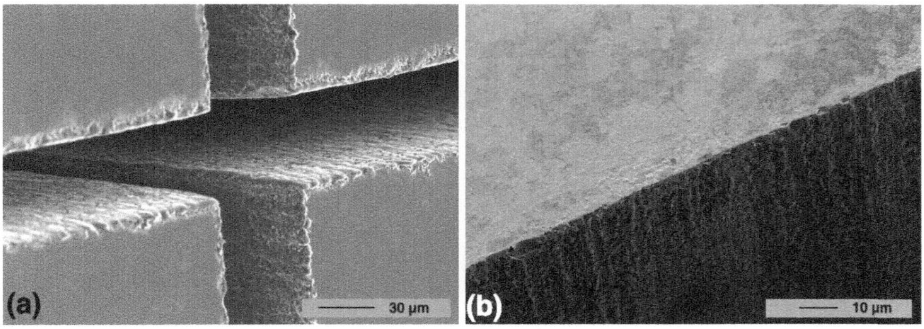

Fig. 8 Silicon wafers cut at 355 nm: **a** crossed cuts in 550 µm wafer (beam entry) and **b** cut in 355 µm wafer (beam entry)

Fig. 9 **a** Silicon wafers cut at 355 nm, section of cut on 550 µm wafer; **b** toothed gear cut with UV-laser

2.4 Mask Projection Ablation of Silicon

2.5-D structures in silicon as applied in micro-fluidics devices, for example, are usually produced by means of a mask projection technique with a homogenized Excimer radiation [7]. The alternative technique by scanning a frequency tripled Nd: YAG-laser beam produces rougher faces because of the localized irradiation. In order to apply this flexible technique nevertheless, quantitative information on the resulting roughness has to be obtained, and ways to minimize this roughness have to be found.

A process parameter that could affect both the ablation depth as well as surface roughness of the final treated area is the number of scans realized, between others. This effect is depicted in Fig. 10. The rectangular pit in Fig. 11 has been obtained by line-wise scanning, with the line direction switched by 40° after every coverage of the area, with nine area scans. The bottom face obtained in this way is fairly well isotropic compared to faces obtained with conventional scanning in two directions. As expected the erosion rate varies essentially inversely with the scanning velocity (Fig. 10). The micro roughness of about 0.8 μm on average is found to be nearly independent of velocity and number of scans but dependent on line spacing. In these investigations, the optimum spacing of 10 μm was chosen.

2.5 Summary

The presented results demonstrate that frequency doubled or tripled solid-state lasers with ultraviolet or green wavelength are suitable for structuring silicon. By judging the obtained accuracy and erosion rate in relation to the effort, the 355 nm is more favourable than the 532 nm wavelength. Although laser structuring does not reach the accuracy of etching, its flexibility and high rate make it a promising

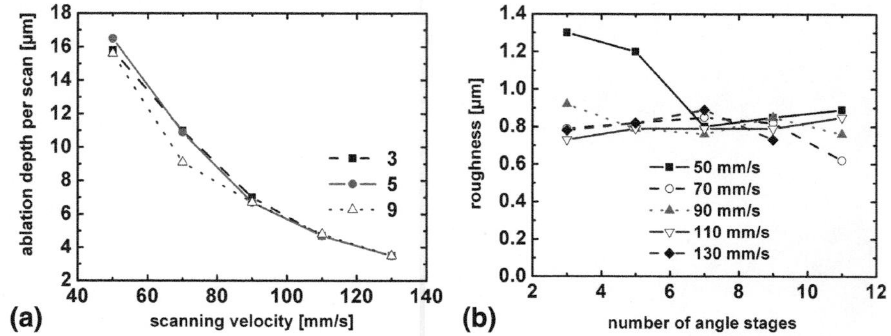

Fig. 10 Processing scanning parameters with changing direction. **a** Average ablation depth per single scan for several numbers of scans; **b** surface roughness for several velocities and numbers of scans

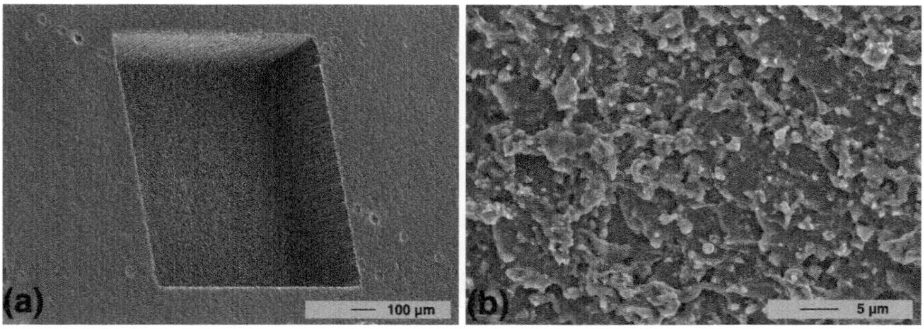

Fig. 11 Erosion of Si (111) face with UV radiation (335 nm): Beam power on sample: 1.1 W, nine scans differing by 40° line direction; scan velocity 70 mm/s. **a** Survey and **b** bottom face magnified

technique, especially in a parameter range where mechanical tooling meets its limits. This technique is expected to be suitable for producing structures in micromechanics, micro-fluidics, and processing of materials for sensors/semiconductor and photovoltaic [8]. In particular, the cutting of irregular-shaped sensor chips might be a promising application.

3 Laser Joining of Low Temperature Cofired Ceramics and Polymers

Low Temperature Cofired Ceramics (LTCC) represent a key technology for the realization of electronic high-frequency systems due to their possible integration into design three-dimensional circuits. Generally, electrical components are produced by screen process. The prepared LTTC foils are stacked, pressed and sintered at high temperature (e.g. 900°C), and utilizing the so called multilayer technology. Three-dimensional ceramic micro structures (like chambers, cavities, channels, through holes) and highly integrated electronic circuits such conductor paths, sensors (e.g. pressure sensor, pH measurement, conductivity, impedance), and actuators (e.g. piezoelectric oscillator) can be realized.

Apart from that, it is also feasible to manufacture large micro-fluidic systems on basis of LTCC ceramic. Those systems are more expensive than equal systems made of polymer due to higher material and processing costs. This technology allows a high flexibility in rapid prototyping of micro reactors and for example online controlling for cell growing with optical microscopy. Compared to a classical stainless steel micro reactor it is possible to integrate electrical sensors or actors—like a impedance measurement system. Another disadvantage of ceramic systems is that the material is not transparent and therefore not suitable for optical analyses. This requirement is necessary to monitor different processes simultaneously as in different areas of biological and medical investigation fields.

To solve this problem, transparent windows (e.g. made of polymer) need to be integrated into the ceramic sensor system.

Different technologies for joining of polymers and ceramic parts are known and established in industrial manufacturing processes. For example, the polymer and ceramic parts can be glued using different kinds of adhesives such organic or inorganic adhesives. The additional material (adhesive) is linking to both joining partners in a chemical (dissolving and mixing) or mechanical (anchoring) way. Another method consists on using mechanical joining techniques like screws or clamps. However, additional features like through holes or notches need to be integrated into the polymer and ceramic sub-assemblies. Using those mechanical joining methods it is difficult to realize gas-proof hermetic joints between the joining partners without inserting an additional sealing.

Laser joining methods have also been employed to join ceramic substrates to other materials (e.g. metallic contacts) [9]. In the last case laser radiation is transmitting through the ceramic joining partner toward the metal joining partner and thus heating both parts as well as the solder. Another common laser joining technique is laser welding of polymers. In this case, both joining partners, made of similar thermoplastic polymers, are irradiated through the first, partly transparent part, while the second part is absorbing the laser radiation. Due to heat conduction processes, the adjacent transparent part is also heated up and thus the material within the contact area is melting and mixing up. After solidification process, the joint has the same properties as the base material.

In the case of ceramic-polymer joining, the materials to be joint have extremely different thermal and physical characteristics. The melting ($\sim$ 100–300°C) and pyrolysis temperatures of common thermoplastic polymers are well below the melting temperatures of common ceramics ($\sim$ 1,000°C). The pyrolysis (thermal destruction) of most common polymers starts below 400°C. Therefore it is impossible to melt and mix both materials for joining purposes. Hence the basic idea of the novel joining technique is to melt the polymer part and pressing the molten material into the roughness and pores of the solid ceramic part. In this way no chemical or material joint is realized but a combination of adhesion and mechanical anchoring between both joining partners.

3.1 The Joining Process

Two different laser sources have been used in recent studies: (i) a standard Nd:YAG solid state laser (Rofin Sinar system RS Marker 100D) and a novel fiber laser system (Ytterbium fiber laser 100C-13-R06), manufactured by SPI. Both systems are emitting continues wave (cw) and pulsed radiation within the near infrared spectrum (1,064 and 1,090 nm, respectively). In addition, different focusing optics with focal lengths between 50 and 300 mm have been used for the joining tests. To realize the relative motion between beam and the joining partners two different systems were utilized. A fixed focusing optic can be combined with a

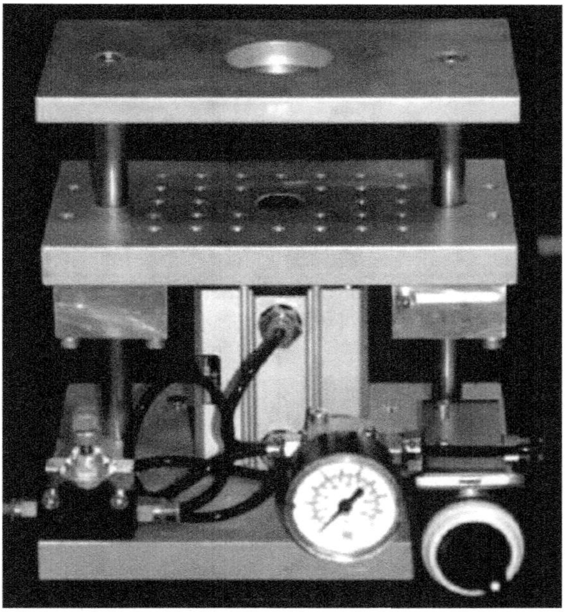

Fig. 12 Pneumatic clamping fixture utilized in the laser joining process

three-axis motion system. The second option is to use a galvanometer scanner for beam deflection. Two pivotable mirrors inside are deflecting and moving the laser beam within a limited marking field. In this case no relative motion of optics and joining parts is needed.

A relevant part of the equipment is the pneumatic clamping fixture depicted in Fig. 12. The joining partners are positioned on top of each other and placed between the upper two plates of the clamp. A cut-out of the upper plate allows the laser beam to reach the joining area. Both parts are pressed against each other with an adjustable pressure.

The clamping fixture has two main functions: realizing a close contact and creating pressure within the molten polymer. A close contact between polymer and ceramic is needed to ensure effective heat conduction between both parts. In addition, it is required that the laser radiation passing through the transparent polymer is not absorbed. When reaching the ceramic part, the radiation is absorbed and converted into heat. Therefore, the heat is conducted from the ceramic to the polymer (the smaller the gap the better the heat transition).

Before joining both materials, the surfaces must be treated. The surface of the ceramic has to be rough or porous enough to ensure that a sufficient density of anchoring points is available for the polymer. The used LTCC parts used for the study show a relative low surface roughness ($R_a \sim 0.5\ \mu m$). In this state the surface is too smooth to realize a strong and reliable bond. In a first step, the LTCC surface was roughened using sandpaper realizing a surface roughness of $R_a = 3.8\ \mu m$ ($R_z = 19\ \mu m$). First successful joints were realized using these mechanically treated surfaces. For more extensive studies, the structuring process

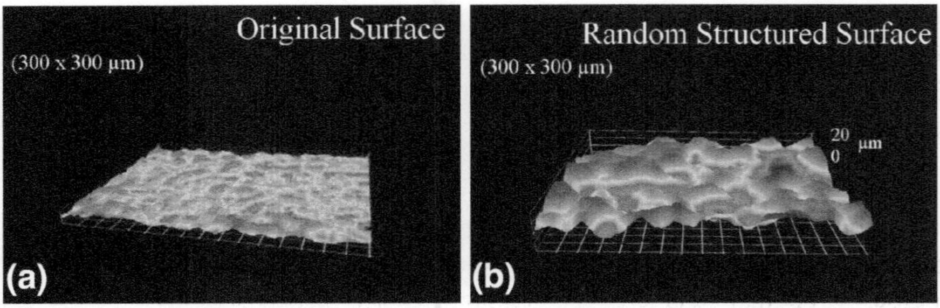

Fig. 13 **a** Unstructured (original) and **b** structured surface with random shot

Table 1 Varied parameters of linear- and dot-shaped structure

Laser pulse frequency (kHz)	Line distance (mm)	Average output power (W)	Structuring velocity (mm/s)
5; 10	0.05; 0.2; 0.7; 1	2; 7; 12;	200; 400; 800; 1,000; 1,500
Laser pulse frequency (kHz)	**Dot density (mm^{-2})**	**Average output power (W)**	**Structuring time per dot (ms)**
10	8.5; 17; 34; 68	15; 20	0.7; 1.0

was accomplished utilizing laser material ablation techniques with pulsed laser radiation [10, 11]. User-defined surface structures (dot- and line-shaped) with different geometrical parameters (such line-distance, density of dots per unit of area, etc.) were fabricated. The programmable laser structuring process offers a high flexibility of the geometries that are generated. Figure 13 shows an example of a typical LTCC surface before and after laser micro structuring. In this case, the Rofin Sinar RS Marker 100 D was used in pulsed "Q-switch mode" with a fixed laser spot size of approximately 100 µm. Therefore the lateral sizes of the single structures (dot- or linear-shaped) were limited to these dimensions (crater diameter or line width). The laser parameters to produce the structures are listed in Table 1.

After preparation of the ceramic surface, the joining partners are placed inside the pneumatic clamping fixture with the polymer part on top, and external pressure is applied. Then, the parts are scanned using the laser radiation (Fig. 14a). If the temperature within the joining area exceeds the fusion temperature of the polymer, a thin layer of molten, high-viscosity polymer is generated. The external pressure is forcing the macromolecules to move into the cavities of the solid ceramic surface (Fig. 14b). When the laser beam has passed or is switched off, the joining area is cooling down due to heat conduction into the bulk material, and the molten polymer solidifies. After the complete part is joined and cooled down the pneumatic clamping fixture can be released. The mechanical anchoring between polymer and the ceramic persists and both parts are joined.

Different laser and process parameters are influencing the property of the joint, e.g. laser output power, joining velocity, intensity, beam diameter, joining pressure

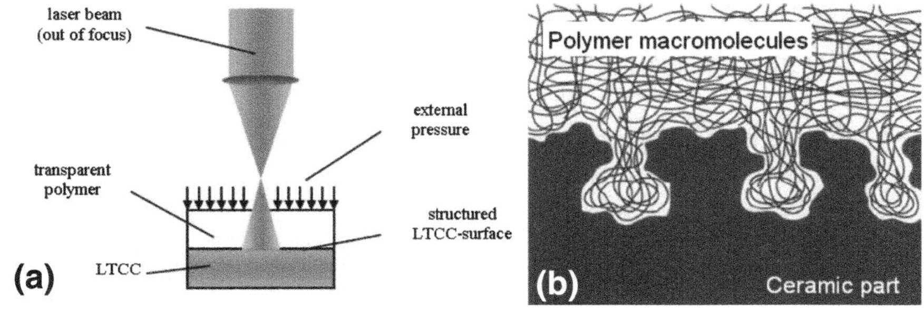

Fig. 14 **a** Principle of the joining process. **b** Schematic of the joined contact area

and surface structure of the LTCC part. For a specific application and material combination the process parameters need to be adjusted. Experimental results showed only minor influence of the external pressure on the joints mechanical strength. The beam intensity is a function of the laser power and the beam diameter. It is defined as the quotient of power and cross-sectional area of the beam. The beam diameter at the joining area can be changed by applying different optics or by changing the focus position (defocusing). By changing the beam diameter (and laser output power) the resulting line width of the joint can be influenced. The beam diameter in focus position approximately defines the minimum line width possible. Several of these parameters are linked to each other, e.g. laser output power, joining velocity and line width. The energy input per unit length is defined as the quotient of power and velocity. Thus the energy input per unit length can be increased by increasing the output power and by decreasing the velocity. This quotient divided by the line width gives the energy input per unit area.

Depending on the energy input per area unit different joining results are emerging. Figure 15 illustrates the changes in the appearance when the energy input is increased. The tensile strength of the joint is influenced as well. In general the tensile strength is rising when increasing the energy input until a maximum strength is reached. When the energy input is too low no joint is generated. Example i shows first reactions of the polymer within the irradiated area. But the amount of molten polymer is too small to flow into the cavities of the ceramic surface. When increasing the energy input (examples ii and iii) the temperature and the amount of molten material increases. The higher temperature reduces the viscosity of the molten material and therefore avails the ability to flow into the ceramic surface. Example iii shows a homogeneous joint without gaps or bubbles. Tensile strength reaches its maximum. Further increasing of the energy input leads to overheating of the molten polymer material (examples iv and v). Small bubbles, a channel in the center and cracks appear within the joint. The tensile strength is decreasing. If the energy input is increased even more, the temperature exceeds the pyrolysis temperature of the polymer resulting in strong thermal damages (example vi).

Concerning to the effect of surface topography of the LTCC parts, the following results were observed. For the smooth untreated LTCC surface, although it was possible to melt the polymer at the joining surface, no strong anchoring between

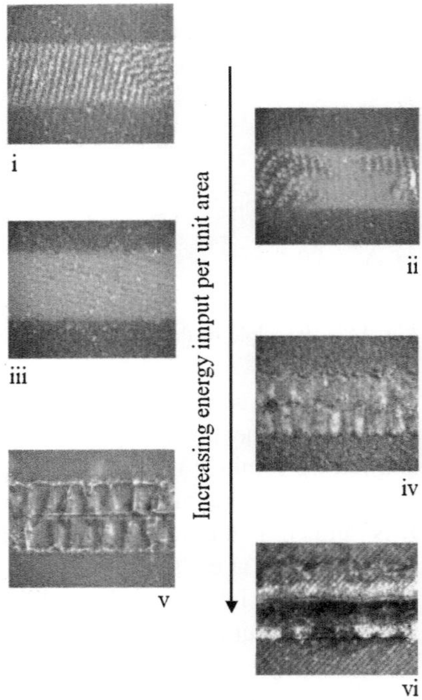

Fig. 15 Different joints depending on energy input (Top-view). The polymer is transparent

both parts was observed. A roughness of $R_a = 0.5$ μm provides an insufficient number and size of anchoring points like cavities to ensure a strong bond. After that, the tensile strength of joined assemblies was tested for a variety of the different patterns and combinations of both patterns. The studies also showed that both structuring patterns are suitable to achieve mechanically strong, reliable and hermetic joints. Both structuring systems reached tensile strengths with similar values. Moreover, recent studies demonstrated that joined parts with a combination of linear and dot-shaped structures achieved the highest tensile strengths (19 N/mm^2). In general, it was observed that the tensile strength is rising when depth and size of the laser generated structures are increasing up to a certain point. When the surface gets too rough, the joint is becoming weaker again. In the last case, more polymer material needs to be molten and pressed into the roughness. But the rough surface decreases heat conduction into the polymer part and therefore hampers the joining process. Thus a compromise has to be found.

3.2 LTCC Based Sensor Example

A LTCC multilayer system consisting of five layers as depicted in Fig. 16 was designed and built. Each layer includes four identical subassemblies of the lab-on-a-chip system. In addition, large circular apertures building up the cell

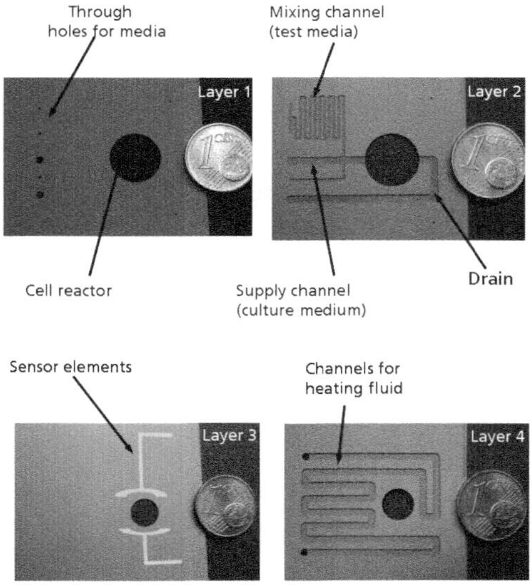

Fig. 16 LTCC multilayer with microfluidic components and impedance sensors

reactors when stacked up were included. At the bottom of layer 4, a large meander like channel is implemented, which is flushed with a tempered fluid to hold a constant temperature within the cell reactor. Above, on layer 3, LTCC based sensors are integrated to measure both impedance and temperature. By drilling different holes, electrical connection within the layers was performed. Impedance measurements are used to monitor changes in cellular growth, e.g. adsorption of cells on a surface. In addition, different media can be fed through the micro channels in layer 2.

This meandrous structure gives the opportunity to mix two different testing liquids or dilution. The through holes of the upper and bottom layers (layers 1 and 5) are used to connect the lab-on-a-chip system to the required supply units and electronic measurement equipment [12].

3.3 Summary

A novel technique permitting the opportunity to join two extremely different materials (polymer and ceramic) without any additional glue or mechanical devices was developed. The joining process is divided into two steps: (i) surface preparation of the ceramic and (ii) laser based partial melting of the polymer part. The joining results are particularly influenced by the energy input per unit area and the shape of the laser structured ceramic surface. Hermetic joints and tensile strength of 19 N/mm^2 were achieved. Based on the technology a lab-on-a-chip system was developed as a first application.

The basic technological principle of the joining technique can also be transferred to join polymers with semiconductors or metals for instance. Promising applications and request for the joining technology are expected in the field of biological and chemical analysis.

The novel joining technique is suitable for a wide range of applications, e.g. combining LTCC based sensors with injection molded polymer microfluidic systems, integrating transparent polymer windows into non-transparent ceramic systems or joining LTCC systems with cheap polymer connectors to realize a connection to supply units. To exemplify the potentials of the developed joining technique a lab-on-a-chip system was developed. The system is designed to analyze the cellular growth within a microbiological reactor under defined conditions.

References

1. Shen, B., Izquierdo, R., Meunier, M.: Laser fabrication of three-dimensional microstructures, cavities and columns. SPIE **2045**, 91–98 (1994)
2. Lorazo, P., Lewis, L.J., Meunier, M.: Picosecond pulse laser ablation of silicon: a molecular-dynamics study. Appl. Surf. Sci. **168**, 276–279 (2000)
3. Bonseo, J., Baudach, S., Krüger, J., Lenzner, M., Wrobel, J.M., Kautek, W.: Femtosecond pulse laser machining of semiconducting materials. In: Conference Digest. "Conference on Laser and Electro-Optics Europe", Piscataway, NJ, USA (2000)
4. Kaspar, J., Luft, A.: Electron microscopic investigation of structural changes in single crystalline silicon induced by short pulse laser drilling. In: Proceedings of the Micro Materials Conference, p. 539, Hrsg. Michel, B., Winkler, T., Druckhaus Dresden GmbH, Dresden (1997)
5. Lenk, A.: Dissertation Komplexe Erfassung der Energie- und Teilchenströme bei der Laserpulse-Anblation von Aluminium. Department of mathematics and natural sciences, Technical University Dresden, Dresden (1996) (in German)
6. Palik, E.D.: Handbook of Optical Constants of Solids II (Palik, Hrsg.). Academic Press, Inc., New York (1997)
7. Crafer, R.C., Oaklay, P.J.: Processing in Manufacturing, p. 163. Chapman & Hall, London (1993)
8. Klotzbach, U., Hauptmann, J.: Potenzial der Mikromaterialbearbeitung mit UV-Lasern für die Photovoltaik; Laserphotonics. Carl Hanser Verlag GmbH & Co. KG, München (2008) (in German)
9. Hesener, H.: Patent WO 0012/256 "method for processing and for joining, especially, for soldering a component or a component arrangement using electromagnetic radiation" (1999)
10. Patent, V.: Franke WO2008025351: "method for producing a bioreactor or lab-on-a-chip system and bioreactors or lab-on-a-chip systems produced therewith"
11. Franke, V., König, A., Klotzbach, U.: Fest gefügt über Materialgrenzen hinweg Mikroproduktion 4/2007. Carl Hanser Verlag GmbH & Co. KG, München (2007)
12. Klotzbach, U., Franke, V., Sonntag, F., Morgenthal, L., Beyer, E.: Requirements and potentialities for bioreactors with LTCC and polymer. Photonics West, San Jose, CA, USA (2007)

Patterning and Optical Properties of Materials at the Nanoscale

Noemí Pérez, Ainara Rodríguez and Santiago M. Olaizola

Abstract The discovery of novel material properties at the nanoscale has aroused a great amount of interest in the fabrication of structures at the sub-micro and nano scales. In this chapter the most promising non-conventional sub-micro and nano fabrication techniques together with the optical characterization techniques which have been developed in recent years to address the novel photonic and plasmonic properties of structured materials are revised.

1 Introduction

Materials structured in the sub-micro and nanoscale present new magnetic [1], chemical [2], mechanical [3] or optical [4] properties. These properties have generated a vast amount of research in the last decade due to their potential use in applications such as human health care and diagnosis [5], sustainable energies [6], textiles [7] or self-cleaning and hydrophobic surfaces [8], among others.

In order to be able to take these applications from the laboratory to industry, it is necessary to develop structuring techniques in the sub-micro and nanoscale with which high resolution at low cost can be achieved. Resulting from this necessity, many scientists have devoted their research efforts to the development of structuring techniques with resolution in the sub-micro and nanoscale with the potential

N. Pérez (✉) · S. M. Olaizola
CEIT and Tecnun (University of Navarra), Manuel de Lardizábal 15, 20018 San Sebastián, Spain
e-mail: nperez@ceit.es

A. Rodríguez
CIC Microgune, Paseo Mikeletegi 48, 20009 San Sebastián, Spain

F. A. Lasagni and A. F. Lasagni (eds.), *Fabrication and Characterization in the Micro-Nano Range*, Advanced Structured Materials, 10,
DOI: 10.1007/978-3-642-17782-8_3,

to scale to industrial fabrication. This chapter is dedicated to the review of different promising techniques paying a special attention to nanosphere lithography (NSL) due to its simplicity and low cost.

Among all the novel properties mentioned above, the optical properties of nanostructured materials are of special interest due to the great amount of applications that derive from them, mainly based on the existence of photonic bandgaps and surface plasmons. These two effects together with the optical characterization techniques that are being developed to characterize materials with these properties are also explained in this chapter. A brief review of the applications of materials with photonic and plasmonic properties is included in the chapter.

2 Structuring Techniques

The most commonly used structuring technique is ultraviolet (UV) lithography with which a resolution of less than 90 nm can be achieved [9]. However, this technique is not widespread outside the semiconductor industry due to its high cost and other techniques as electron [10] or ion beam lithography [11, 12] are used in research laboratories to precisely structure different types of materials. They offer high resolutions but both are time consuming techniques, making them unsuitable for mass production. There are some non-conventional techniques as laser interference lithography (LIL) [13], nano-imprint lithography (NIL) [14] or nanosphere lithography [15] which have the potential to scale to industrial production at a lower cost than those mentioned above. In this section, these non-conventional techniques are briefly introduced.

2.1 Nano Imprint Lithography

Nano imprint lithography (NIL) is a structuring technique which uses a hard mould to obtain the desired structure. The mould is pressed against a resist or a polymer, creating a pattern which can be transferred to the substrate by a conventional etching technique as, for example, reactive ion etching (RIE).

There are two different ways in which the pattern can be transferred to the material. The most widespread is thermoplastic NIL [14]. In this approach, after pressing the mould against the polymeric material, it is heated until it can flow and fill in the geometries in the hard mould. When the material is cooled down, the mould is released from it and the material remains patterned. The second approach is UV-assisted NIL [16], in which UV curable resists and transparent moulds are used. In this case, after pressing the mould against the resist, it is cured with UV light that passes through the mould. The advantage of this technique with respect to thermoplastic NIL is that there is no need for the high temperatures required in

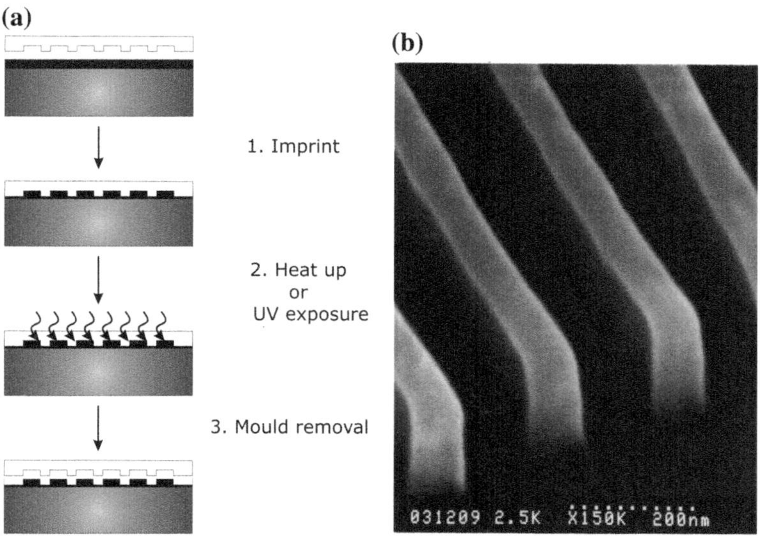

Fig. 1 **a** Nano imprint lithographic full process. **b** *70 nm wide fringes patterned by thermoplastic NIL in PMMA. *Reprinted with permission from Chou et al. [14], Copyright [1996], AVS—The Science & Technology Society

the latter. Figure 1 shows a diagram of the NIL process together with the result of a thermoplastic NIL process on a PMMA substrate.

The resolution of NIL depends on the moulds which are frequently fabricated by electron beam lithography, making them quite expensive. Moreover, these moulds have to be replaced after less than one hundred uses, which increases the cost of the technique. The highlight of this technique is the possibility of patterning large areas in a short time and with any structure which can be engraved in the mould.

2.2 *Laser Interference Lithography*

Laser interference lithography (LIL) is based on the structuring of materials with the resulting periodical pattern from the interference of two or more coherent laser beams. This patterning can be done by direct ablation of the materials [17] or by a complete lithographic process, combining LIL with conventional etching or deposition techniques.

LIL is a maskless technique with the potential to structure large areas. Its resolution is only limited by the wavelength of the laser source, so the minimum obtainable dimensions can be scaled down just by choosing laser beams with shorter wavelengths. Although this technique can be used in a full lithographic process, one of its main advantages resides in the use of high-energy laser beams which provide the capacity for the direct ablation of materials.

As LIL is explained in depth in Chap. 1, it will not be treated in more detail in this section. However, it is worth mentioning the potential of this technique to scale to industrial applications. In fact, there are already commercial systems which allow the patterning of materials with two laser beams in a Lloyd's mirror configuration. Nevertheless, the use of three and four beams has the technical difficulty of the laborious manual alignment, which holds this technique back from scaling to industrial applications. Recent inventions which allow the automatic alignment of the system [18] bring the LIL technique closer to industry.

2.3 Nanosphere Self-Assembly Techniques

The self-assembly of colloidal nanospheres in order to obtain the so called artificial opals [19] has been studied in the last 20 years in order to obtain photonic crystals for infrared and visible light. These opals do not meet, in general, the necessary requirements to obtain complete photonic bandgaps (see Sect. 3.1) due to two main reasons [20]. The first one is the filling factor, which must be about 20% for compact hexagonal structures, just the opposite of artificial opals. The second one is the refractive index contrast, which is not enough taking into account that in most cases silica or polystyrene spheres are used, which have refractive indexes about 1.5. As a result, many research efforts have been dedicated to the inversion of the original structures to obtain inverse opals [21, 22]. Nanosphere lithography starts from this self-assembly idea with the aim of obtaining sub-micro and nanometric structures in large areas with low cost, usually with order in two dimensions.

Following, the different steps which lead to the fabrication of three dimensional inverse opals are presented together with the evolution of this technique towards the so-called nanosphere lithography.

2.3.1 Three-Dimensional Inverse Opals

The fabrication of inverse opals is considered itself as a structuring technique with which three dimensional structures can be obtained. The whole process consists of three different steps, which are summarized in Fig. 2.

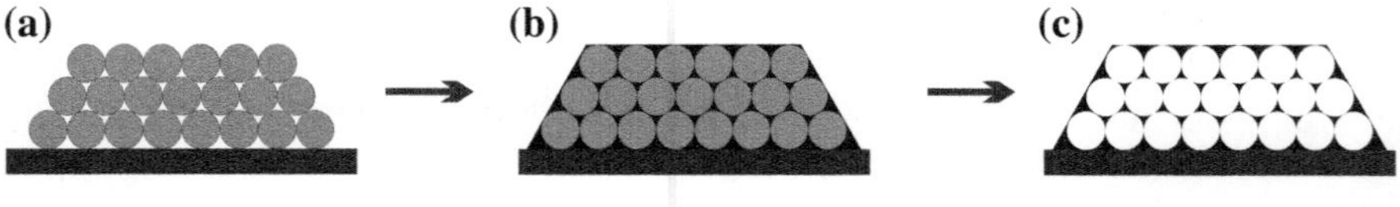

Fig. 2 Scheme of the fabrication process of an inverse opal. The first step is the self-assembly of the colloidal spheres to form the opal (**a**), the second step is the infiltration of the opal matrix with the desired material (**b**), the third step is the removal of the opal matrix (**c**)

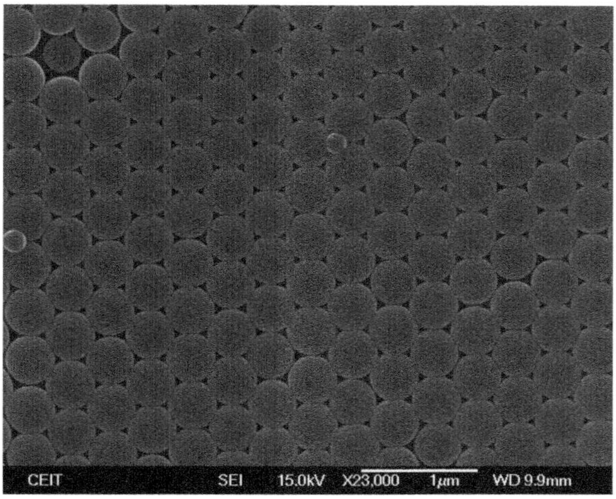

Fig. 3 Opal fabricated by vertical deposition with 419-nm polystyrene spheres

The first step is the self-assembly of colloidal nanospheres to obtain the three-dimensional matrix with which the required material will be structured. The self-assembly issue itself has originated a vast amount of research and, in this context, techniques as sedimentation [23, 24], vertical deposition [19], spin coating [25] or electrophoretic deposition [26] have been developed. In all cases, the obtained artificial opal has a compact hexagonal structure with large scale order in three dimensions. A polystyrene artificial opal is shown in Fig. 3.

One of the main issues of this technique is the substrate hydrophilicity, which is crucial for the correct adhesion of the nanospheres. This property is used by some authors to obtain opals only in certain regions by defining hydrophobic and hydrophilic areas [27], as can be seen in Fig. 4.

Once the opal matrix is obtained, the second step is the infiltration of its voids with the desired material. Many different infiltration techniques are used depending on the type of material. Semiconductors such as germanium [28] and silicon [29, 30], are infiltrated by chemical vapor deposition. The sol–gel deposition technique is used for the infiltration of silica [31, 32], alumina [33] or barium titanate [34]. Since the first infiltrated opals by electrodeposition [35], it has become the technique of choice for the infiltration of different metals such as

Fig. 4 Selective formation of opals in hydrophilic areas. Reprinted figure with permission from Fustin et al. [27]. Copyright Wiley-VCH Verlag GmbH & Co. KGaA

Fig. 5 Silver inverse opal fabricated from a polystyrene opal by electrodeposition and chemical removal of the 419-nm polystyrene spheres with toluene

gold [36], silver [37] or platinum, palladium and cobalt [38]. Even the infiltration of the opal matrixes with silica nanoparticles and their subsequent sintering has been reported by some authors [39, 40]. More in-depth reviews on the fabrication of inverse opals by the different techniques mentioned above can be found in bibliography [41–43].

The last step of the fabrication process is the removal of the opal matrix to obtain the structured material. The inversion methods used are different depending on the material which composes the spheres. If the spheres are made of silica, their elimination is carried out by chemically dissolving the silica with hydrofluoric acid [44]. If the spheres are polymeric, commonly polystyrene or Polymethyl methacrylate (PMMA), they can be removed either by chemical or physical methods. As for the chemical methods, polymeric spheres are dissolved with toluene [35] or tetrahydrofuran (THF) [45]. The physical method implies the calcination of the polymeric spheres at high temperatures [46]. A silver inverse opal prepared by electrodeposition can be seen in Fig. 5.

2.3.2 Nanosphere Lithography

Although only three dimensional structures are explained above, two dimensional structures can also be obtained if a one-layer nanosphere matrix is formed in the first step. There are many self-assembly techniques with which a single layer of spheres can be obtained; however, spin coating is usually the technique of choice for many authors due to its simplicity and repeatability [47, 48].

Once this nanosphere layer is formed, it can be used as a mask in two different approaches for subsequent deposition or etching steps, leading to different structures. These two approaches conform what it is known as nanosphere lithography, first called natural lithography [49]. Due to the spreading of NSL, some authors also refer to it when dealing with the fabrication of three-dimensional inverse opals.

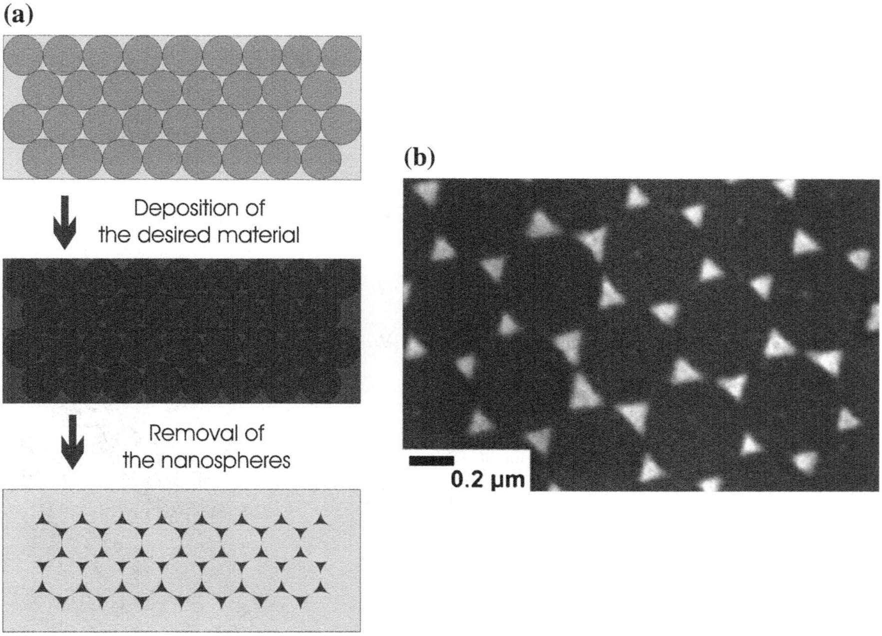

Fig. 6 **a** Nanosphere lithographic process for the fabrication of non connected nanoparticles. **b** *Gold nanoparticles fabricated by NSL. *Reprinted figure with permission from Canpean and Astilean [55], Copyright (2009), Elsevier

The first NSL approach comprises the filling and the removal of the nanosphere structure just as explained above. This approach, which is sketched in Fig. 6, gives rise to hexagonally ordered triangle-shaped nanoparticles. If polystyrene spheres are used their size can be decreased by undergoing an oxygen RIE process [15, 47, 50, 51]. In that way, connected structures of voids are obtained (see Fig. 7) and the diameter of the voids can be modified by changing the diameter of the spheres.

A modification of the original NSL technique known as, angle resolved nanosphere lithography, allows controlling the size, shape and spacing of the nanoparticles [52]. This approach produces a varied amount of nanostructures by controlling the angle between the substrate that supports the nanosphere matrix and the direction of the material deposition.

In the second approach the nanosphere ordered matrix is used as an etching mask. The obtained structures in this case are nanopillars and their diameter can be modified by two different means. One is the thinning of the polystyrene spheres prior to etching as explained above. The other has already been used with silicon [53] and consists in the oxidation of the pillars in order to obtain a thin layer of silicon oxide on their surface. This oxidized layer is removed by hydrofluoric acid and, as a result, the pillars get thinner. The thickness of the silicon oxide layer and, thus, the final thickness of the nanopillars, can be controlled by the oxidation time.

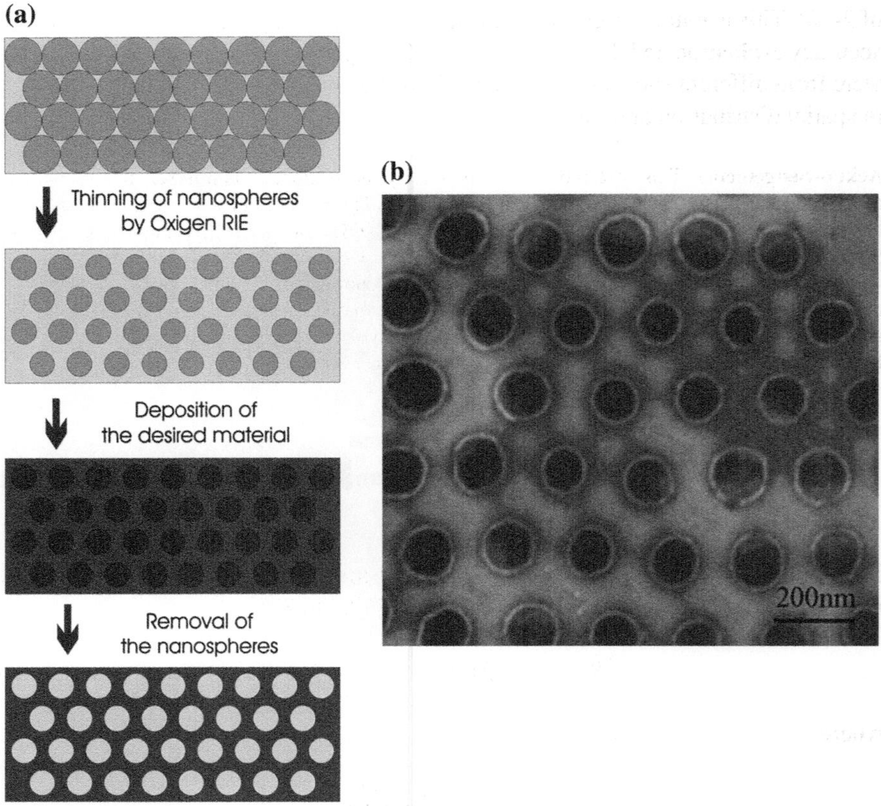

Fig. 7 **a** Nanosphere lithographic process with a RIE step to reduce the size of the spheres. **b** *Gold connected structure fabricated by NSL. *Reprinted figure with permission from Zhang et al. [51], Copyright (2008), Elsevier

Some structures fabricated with this technique together with a diagram summarizing this NSL approach are shown in Fig. 8.

The materials patterned with nanosphere lithography are very varied: silicon [53, 54] and silicon oxide [47] nanopillars have been fabricated to be used as moulds for nano imprint lithography and different metals as gold [15, 51, 55], silver [15] or platinum [50] have been nanostructured for their potential plasmonic properties.

3 Optical Properties of Structured Materials

In this section, the concepts of photonic crystal and surface plasmon are introduced as well as different techniques used to characterize structures with these optical properties. Finally, some of the main applications of such nanostructured materials are presented.

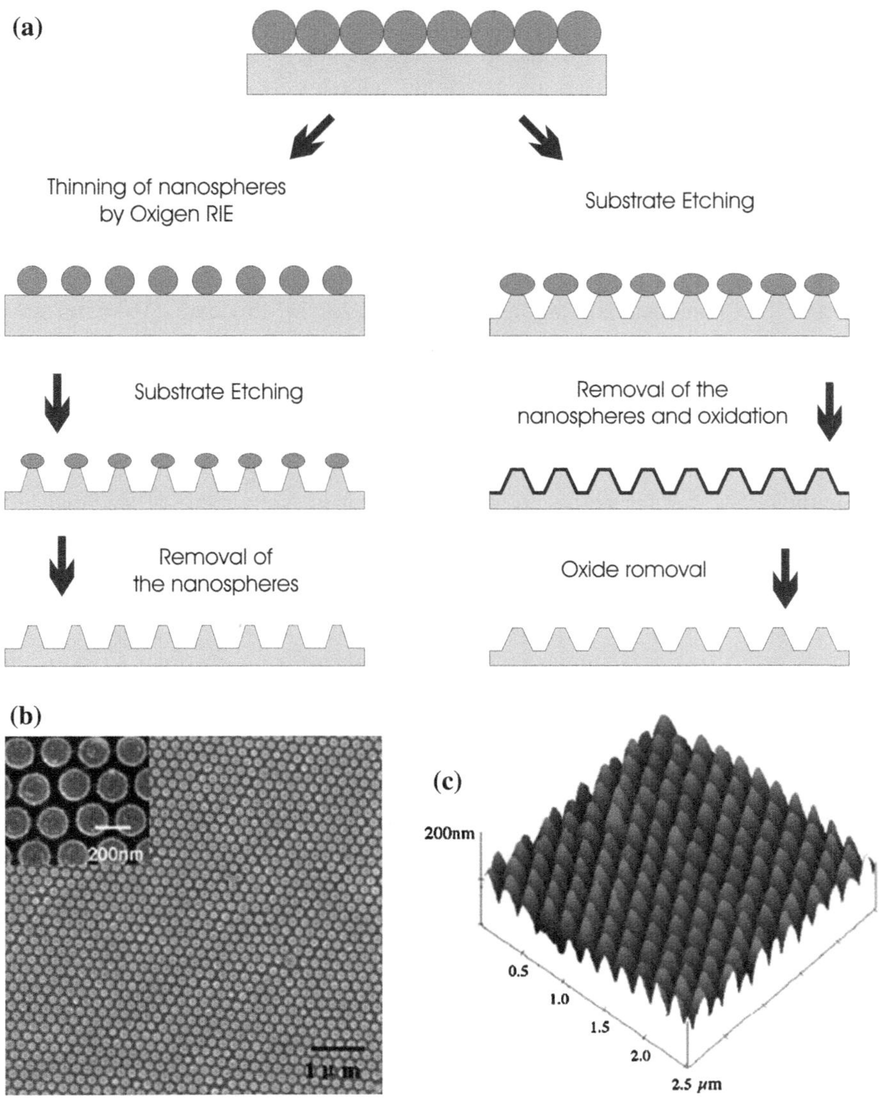

Fig. 8 **a** Nanosphere lithographic process for the fabrication of pillars. **b** *SEM micrograph and **c** *AFM topography of an array of silicon nanopillars fabricated by NSL and thinned by the oxidation approach. *Reprinted figure with permission from Li et al. [53], Copyright (2007), Elsevier

3.1 Photonic Crystals

The complete control that some crystalline materials such as semiconductors provide over electrons is a well-known property of this type of materials, which are essentially periodical arrangements of atoms. This periodicity is a key characteristic for the formation of electronic bandgaps in semiconductors.

The analogous optical structures to semiconductors are photonic crystals, which are periodical arrangements of materials with different dielectric constants (refractive indexes). Photonic crystals present photonic bandgaps, defined as energies with which photons cannot propagate inside them. Such complex systems cannot be found in nature and thus, they have to be fabricated artificially. The most simple photonic crystals are one-dimensional, meaning they posses one-dimensional periodicity and, consequently, one-dimensional photonic bandgaps. Conversely to two and three-dimensional photonic crystals, these one-dimensional structures are quite easily fabricated by alternating deposition of thin layers of two different materials [56].

From the definition of photonic bandgap, it can be inferred that a photonic crystal will reflect all the light which affects it if the energy of the light has values within its bandgap. If a beam of light with certain energy is reflected from a photonic crystal for any polarization of the beam and at any angle of incidence it is said that the photonic crystal possesses a complete photonic bandgap (CPBG) at that energy. Such structures are a challenge to fabricate due to the necessity of periodicity in the three dimensions of space. The first structure with a CPBG was fabricated in 1991 [57] by drilling holes in a slab of dielectric with a certain orientation. It was first fabricated in the centimeter scale in order to study the propagation of microwaves inside it. The result was a diamond-like structure with a CPBG between the second and the third bands. Three years later another structure with a complete photonic band gap was fabricated [58]. In this case, it was a woodpile structure built as a layer-by-layer structure of horizontal columns. It was built by alternating standard deposition and photolithographic processes, which represents an advantage in the fabrication compared to the structures explained before. Ten years later another structure with a photonic bandgap was fabricated by alternating rod and hole layers [59]. Each double layer (rod-hole) was fabricated in one process cycle, consisting of four subsequent steps of deposition, lithography, etching and planarization. All the approaches explained to the moment comprise many different steps which make the fabrication processes of photonic crystals with complete photonic bandgaps time consuming and not suitable for mass production. Using de inverse opal approach to create three-dimensional ordered structures, silicon [29] and gold [60] photonic crystals with CPBGs were obtained and, therefore, it was demonstrated that NSL is a competitive technique for the fabrication of three-dimensional photonic structures.

3.1.1 Characterization of Photonic Crystals

When characterizing a nanostructured material with potential photonic properties, it must first be confirmed the existence or not of a photonic bandgap and then it must be verified if the photonic bandgap is complete or not. As a beam of light with its energy within the bandgap is reflected from a photonic crystal for any angle of incidence and polarization, photonic crystals can be characterized by relatively simple optical measurements. The most commonly used technique for the characterization of the optical response of photonic crystals is the measurement of their reflectance at different angles of incidence [29, 61–64], often known as angled-resolved reflectivity.

Although there are many different setups used for this type of measurements, all of them include a white light source, a spectrometer and a goniometer to which the sample is attached. The goniometer is used to adjust the angle of incidence and the angle of reflectance at which the measurements are taken, which must be the same. In the example set up shown in Fig. 9 the goniometer is attached to a translation stage which allows the movement of the sample in three different axes. There is also a Glan–Thomson polarizer which permits the study of the polarization dependence of the optical response of the nanostructured material. The light reflected from the sample is imaged in a screen which has a 500 μm pinhole in it and, thus, only a small area of the sample is characterized each time. The reflected spectrum from the selected area enters a spectrograph which decomposes the reflected light into its wavelength components and this spectrum is imaged with a CCD camera. Different spectra are measured for several angles of incidence thus characterizing the optical response of the selected nanostructured material.

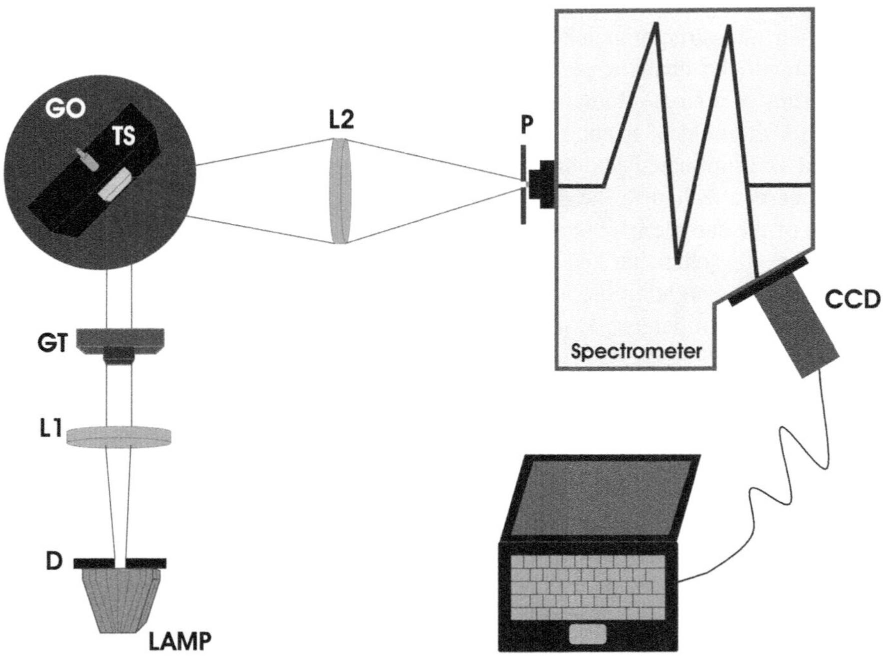

Fig. 9 Scheme of an angle-resolved reflectivity setup, where D is an iris diaphragm, L1 and L2 are 50-mm focal lenses, GT is a Glan–Thompson polarizer, TS is a translation stage, GO is a goniometer and P is a screen with a 500-μm pinhole on it

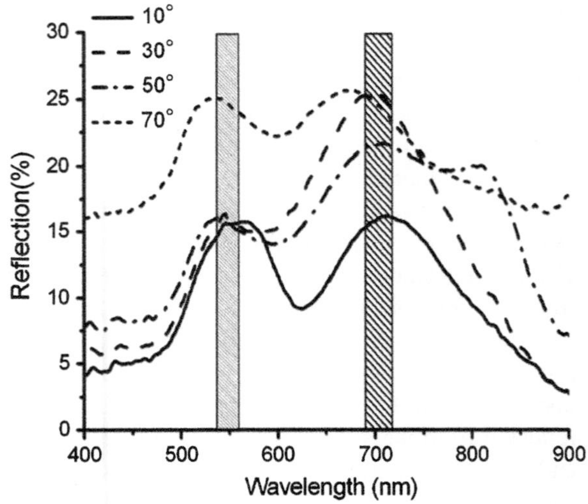

Fig. 10 Reflectance measurements of a gold structure with a CPBG. The reflectance peaks overlap for any angle of incidence. Reprinted figure with permission from Kuo and Lu [60], Copyright (2008), American Institute of Physics

If the characterized material presents reflectance peaks which overlap for every angle of incidence it can be concluded that the material possesses a photonic bandgap. When this occurs for unpolarized light, the photonic bandgap is then complete. The reflectance measurements of a gold inverse opal with a complete photonic bandgap are shown in Fig. 10.

When nanostructured materials with crystalline domains of different orientations are to be characterized, it may be interesting to characterize the optical properties of a single domain. In this case, single domain spectroscopy is carried out with the help of a microscope [65]. A portion of the material to be characterized is illuminated and imaged with the help of a microscope. An external eyepiece is attached to the output port of the microscope with which a reflected image of the sample is obtained at the entrance slits of a spectrometer. These slits determine the characterized area of the photonic crystal and permit to select a single domain. In addition, a scanning of small areas of a photonic crystal can be done in order to determine the homogeneity of its optical properties.

3.1.2 Applications of Photonic Crystals

There are two determining factors for the existence and the position of a photonic bandgap. One is the refractive index contrast between the materials which form the photonic structure and the other is its geometry. If any of these two parameters changes in the presence of a certain substance, they could be used as sensors. Based on the change in the refractive index contrast, different sensors as a hazardous organics in water sensor [66] or a protein sensor [67] have been developed. Using the geometry modification approach, an innovative glucose sensor has been developed [68]. It consists in a polymerized crystalline colloidal array with a photonic bandgap which contains molecular recognition element. In the presence of glucose the restoring force of the polymeric array is increased, decreasing the size of the photonic crystal and shifting the respective reflectance peak towards shorter wavelengths.

However, the most widespread application of photonic crystals is their use in optical circuits as waveguides [69–71]. For this type of applications, only two-dimensional photonic crystals are needed, which are mainly fabricated by electron beam lithography. Other optical components as mirrors or splitters have also been proposed [72]. Moreover, the use of photonic crystals is being studied for the tailoring of the emission beam patterns of lasers [73].

3.2 Surface Plasmons

There is an interesting effect that happens at the interface between a metal and a dielectric which was first studied by Ritchie and co-workers in 1957 [74]. This effect is the generation of the so-called surface plasmons which are waves propagating along the surface of the metal [4, 75]. In brief, the electrons inside metals are present in the form of an electronic plasma which can oscillate in the presence of an electromagnetic wave. In this case, it is light what induces an interaction with the electrons within the metal at a certain resonant frequency generating waves which are frequently called surface plasmon polaritons (SPPs) due to their hybrid nature (photon–plasmon). However, most authors refer to them simply as surface plasmons (SPs). Although SPs are propagating waves along the metal surface, the field perpendicular to the surface is evanescent, meaning that its amplitude decreases exponentially with increasing distance from the surface. As a result, surface plasmons cannot propagate away from the surface. It is worth mentioning that the momentum of surface plasmons is greater than the momentum of a free-space photon of the same frequency. Thus, in order to excite a surface plasmon, it is needed to provide the difference in the momentum between the incident photons and plasmons. This has been classically achieved with the use of a prism as depicted in Fig. 11. When a white light beam hits the metallic surface (gold, in this case) through a prism at a certain angle of incidence, light is coupled to the metallic plasma, generating a surface plasmon. The generated SPs are very sensitive to any changes in the dielectric below the metallic layer and, for this reason, SPs have been widely studied in recent years for their unique sensing properties as it will be seen further in this section.

The behavior of metallic nanostructured materials presents significant differences when comparing to flat metallic surfaces. The unusual coupling of light to metallic nanoparticles was first noticed by Faraday in 1857 [76], many years before Ritchie and coworkers [77] revealed the existence of plasmons in structured metals and their coupling with light. In brief, light incident on a metallic patterned surface can scatter to higher-momentum states and excite a surface plasmon at a certain resonant frequency with no need of a prism. In the case of metallic nanoparticles, the value of this resonant frequency depends on their shape, size and

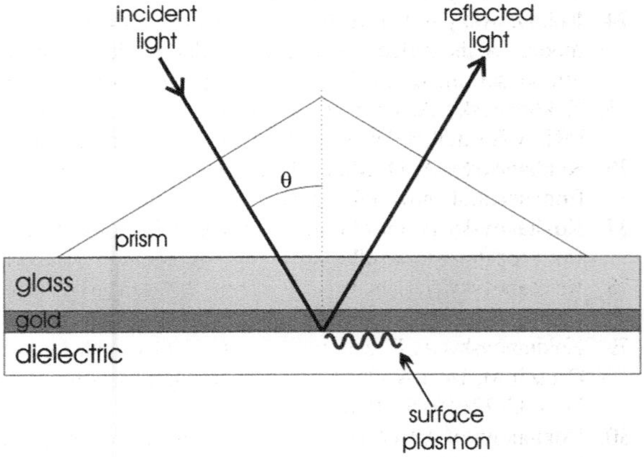

Fig. 11 Prism coupling of light and excitation of a surface plasmon at the interface between gold and a dielectric

composition. Similarly to SPs generated in metallic flat surfaces, they are very sensitive to the properties of the material surrounding the metallic nanoparticles. But there is a difference between these two types of interactions. In the case of flat metallic surfaces the created surface plasmons can propagate meanwhile the SPs generated in metallic nanoparticles are localized and, thus, this phenomenon is known as localized surface plasmon resonance (LSPR). The scientific and technological interest of this plasmonic effect has caused that most of the efforts to fabricate metallic nanoparticles with NSL are with the aim to produce materials with localized surface plasmons (LSPs).

Not only nanoparticles enhance the coupling of light with plasmons, but also any structured metals with sub-wavelength features. Ebbesen and co-workers [78] presented, in 1998, the first observation of the enhancement of light transmission through a patterned metallic layer due to the coupling between light and plasmons. They observed that a gold layer patterned with an array of sub-wavelength holes transmitted more light than expected, concluding that light was coupled to the plasmons at the metal surface. As the structured metallic film was thin enough the plasmons of both sides got coupled and, as a result, the transmission of light was intensified. In this case, plasmons are not localized and are able to propagate along the patterned metallic surface. Although the theory behind this event was explained in 2001 [79] many authors still study this phenomenon today due to the scientific and technological implications of this finding [80].

3.2.1 Characterization of Surface Plasmons in Nanostructured Materials

Similarly to what happens with photonic crystals, although the physical phenomena that govern the generation of surface plasmons are quite complex, the existence of this light–metal interaction can be confirmed by relatively simple optical measurements.

Roughly, when an optical wave excites a SP in a metallic nanoparticle, there is an intense wavelength-selective absorption and scattering of energy from the nanoparticle. This energy can be measured in three different ways: by measuring the extinction spectrum of the nanoparticles, by measuring their reflectance spectrum or by imaging the nanoparticles taking advantage of the light that they scatter.

The most common way of characterizing the LSPR response of a metallic nanoparticle (or an array of metallic nanoparticles) is the measurement of their extinction spectrum [81–85], which is defined as the sum of scattering and absorption. In particular, extinction is defined as $-\log(I_t/I_0)$, where I_0 and I_t are the incident and transmitted intensities of light, respectively. The absorption and scattering of light in the LSPR phenomena happens mainly in the ultraviolet and visible region of the spectrum and thus, UV–vis extinction spectroscopy is carried out in order to characterize the extinction spectrum of the nanoparticles. It consists in the measurement of the intensity of light (commonly white light) which is transmitted through the nanoparticle or nanoparticle array which is to be characterized. This is commonly carried out with a simple set-up containing an unpolarized white light source and a spectrometer. Obviously, this technique can only be used with transparent substrates. The parameter that characterizes the extinction spectrum is the wavelength of maximum extinction (λ_{max}) which is commonly located by calculating the zero-crossing point of the first derivative [82]. The extinction spectrum of a silver nanostructure fabricated by laser interference lithography is shown in Fig. 12. The nanostructure is first characterized in air atmosphere and then it is surrounded by alcohol. This produces a shift in the wavelength of maximum extinction, as can be seen in the figure.

If the nanoparticles are patterned on an opaque substrate, the extinction coefficient cannot be measured and the LSPR is characterized by the reflectance spectrum. As a great amount of light is absorbed when a LSP is excited, the reflectance spectrum presents an absorption peak at a certain wavelength (λ_{min}), which is the parameter used to characterize the LSPR in the reflectance mode. The reflectance is measured in a similar way to that previously explained for photonic crystals (see Sect. 3.1.1).

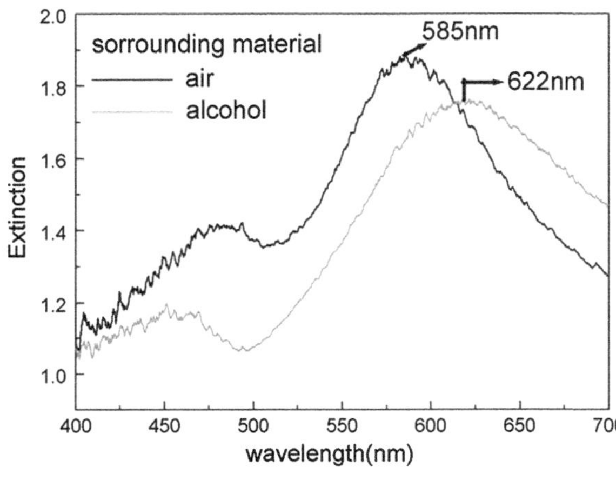

Fig. 12 Extinction spectrum of a LIL fabricated silver nanostructure surrounded by two different media: air and alcohol. Reprinted figure with permission from Li et al. [86]. Copyright (2008), Elsevier

The last approach to LSPR characterization is the imaging of the nanoparticles taking advantage of the light they scatter. However, these measurements are more complicated that those explained above as microscopes and dark field illuminators are needed [87]. From these measurements, even the resonant Rayleigh scattering spectrum can be measured which presents scattered intensity peaks [88]. The wavelength position of these peaks is also used to characterize the LSPR response of nanoparticles and nanoparticle arrays.

3.2.2 Applications of Surface Plasmons

As has been mentioned before, the properties of the surface plasmons excited in nanoparticles, mainly the position of the maximum extinction (λ_{max}) and the reflectance peak (λ_{min}), are highly sensitive to variations in the properties of the surrounding dielectric material and, thus, the variations in the medium can be measured by monitoring the changes in the mentioned parameters. In particular, an increase in the dielectric constant or in the thickness of the surrounding material produces a red-shift in the position of the maximum extinction [89]. Thus, metallic nanoparticles (and nanoparticle arrays) can be used as transducers which convert small changes in refractive index near their surface into a measurable wavelength shift response [85]. This approach has been widely studied [90, 91], mainly for biosensors [92–96] and gas sensing devices [97–102]. It must be taken into account that other characteristics as the size and shape of the nanoparticles change the response of the LSPs and, thus, the right election of these parameters may lead to a better response of the plasmonic sensor [88].

The extraordinary propagation properties of SPs in structured materials have been suggested for applications in communications such as the fabrication of photonic circuits [4, 103] and devices [91, 104]. Surface plasmons have also been suggested in order to increase the efficiency of light emitting devices [105, 106] or solar cells [107]; and even for the fabrication of lasers [108].

Highly amplified local electromagnetic fields are generated at the surfaces of nanoparticles due to the absorption and scattering of light that the excitation of surface plasmons gives rise to. Thus, this type of nanostructured materials is also used to amplify surface-enhanced Raman scattering (SERS) [109].

4 Conclusions

In this chapter different structuring and optical characterization techniques in the sub-micro and nanoscale have been presented.

There are some lithographic techniques such as UV, electron beam and ion beam lithography with which nanometric resolution can be achieved but with the drawback of very high costs. With some non-conventional techniques such as nano imprint lithography, laser interference lithography and nanosphere lithography similar resolutions can be achieved but in a cost-effective way.

Nanosphere lithography has evolved from the fabrication of inverse opals and permits the fabrication of different structures such as nanoparticles or nanopillars.

Materials patterned with features in the sub-micro and nanoscale present different outstanding optical properties. The existence of photonic bandgaps in periodically patterned materials and the possibility of exciting surface plasmons in metallic nanoparticles in a much easier way than in flat metals have generated a vast amount of research in the last 20 years. This is due to the potential applications of these two phenomena in different fields mainly related to sensing.

Although the photonic and plasmonic properties of structured materials are complex in their basis, they can be characterized with relatively simple optical measurements which have been explained in this chapter.

References

1. Thomas, G., Hutten, A.: Characterization of nano-magnetic structures. Nanostruct. Mater. **9**, 271–280 (1997)
2. Kung, H.H., Kung, M.C.: Nanotechnology: applications and potentials for heterogeneous catalysis. Catal. Today **97**, 219–224 (2004)
3. Ovid'ko, I.A.: Deformation of nanostructures. Science **295**, 2386 (2002)
4. Atwater, H.A.: The promise of plasmonics. Sci. Am. **296**(4), 56–63 (2007)
5. Sahoo, S.K., Parveen, S., Panda, J.J.: The present and future of nanotechnology in human health care. Nanomed. Nanotechnol. Biol. Med. **3**, 20–31 (2007)
6. Serrano, E., Rus, G., García-Martínez, J.: Nanotechnology for sustainable energy. Renew. Sustain. Energy Rev. **13**, 2373–2384 (2009)
7. Brown, P., Stevens, K.: Nanofibers and Nanotechnology in Textiles. Woodhead, Cambridge (2007)
8. Bhushan, B., Jung, Y.C.: Natural and biomimetic artificial surfaces for superhydrophobicity, self-cleaning, low adhesion, and drag reduction. Prog. Mater. Sci. doi: 10.1016/j.pmatsci.2010.04.003
9. ITRS: International technology roadmap for semiconductors (2007)
10. Rius, G.: Electron Beam Lithography for Nanofabrication. PhD thesis, Universidad Autónoma de Barcelona (2008)
11. Reyntjens, S., Puers, R.: A review of focused ion beam applications in microsystem technology. J. Micromech. Microeng. **11**, 287–300 (2001)
12. Matsui, S., Ochiai, Y.: Focused ion beam applications to solid state devices. Nanotechnology **7**, 247–258 (1996)
13. Kim, D.Y., Tripathy, S.K., Li, L., Kumar, J.: Laser-induced holographic surface relief gratings on nonlinear optical polymer films. Appl. Phys. Lett. **66**, 1166–1168 (1995)
14. Chou, S.Y., Krauss, P.R., Renstrom, P.J.: Nanoimprint lithography. J. Vacuum Sci. Technol. B **14**, 4129–4133 (1996)
15. Hulteen, J.C., Van Duyne, R.P.: Nanosphere lithography: a materials general fabrication process for periodic particle array surfaces. J. Vac. Sci. Technol. A **13**, 1553–1558 (1995)
16. Haisma, J., Verheijen, M., van den Heuvel, K., van den Berg, J.: Mold-assisted nanolithography: a process for reliable pattern replication. J. Vac. Sci. Technol. B **14**, 4124–4128 (1996)

17. Lasagni, A., Holzapfel, C., Mücklich, F.: Production of two-dimensional periodical structures by laser interference irradiation on bi-layered metallic thin films. Appl. Surface Sci. **253**, 1555–1560 (2006)
18. Rodriguez, A., Echeverria, M., Ellman, M., Perez, N., Verevkin, Y.K., Peng, C.S., Berthou, T., Wang, Z., Ayerdi, I., Savall, J., Olaizola, S.M.: Laser interference lithography for nanoscale structuring of materials: from laboratory to industry. Microelectron. Eng. **86**, 937–940 (2009)
19. Jiang, P., Bertone, J., Hwang, K., Colvin, V.: Single-crystal colloidal multilayers of controlled thickness. Chem. Mater. **11**, 2132–2140 (1999)
20. Sözüer, H.S., Haus, J.W., Inguva, R.: Photonic bands: convergence problems with the plane-wave method. Phys. Rev. B **45**, 13962–13972 (1992)
21. Norris, D.J., Vlasov, Y.A.: Chemical approaches to three-dimensional semiconductor photonic crystals. Adv. Mater. **13**, 371–376 (2001)
22. Galisteo López, J.F., García-Santamaría, F., Golmayo, D., Juárez, B., López, C., Palacios, E.: Self-assembly approach to optical metamaterials. J. Opt. A Pure Appl. Opt. **7**, S244–S254 (2005)
23. Xia, Y., Gates, B., Yin, Y., Lu, Y.: Monodispersed colloidal spheres: old materials with new applications. Adv. Mater. **12**, 693–713 (2000)
24. Vickreva, O., Kalinina, O., Kumacheva, E.: Colloid crystal growth under oscillatory shear. Adv. Mater. **12**, 110–112 (2000)
25. Mihi, A., Ocaña, M., Míguez, H.: Oriented colloidal-crystal thin films by spin-coating microspheres dispersed in volatile media. Adv. Mater. **18**, 2244–2249 (2006)
26. Trau, M., Saville, D.A., Aksay, I.A.: Field-induced layering of colloidal crystals. Science **272**, 706–709 (1996)
27. Fustin, C.A., Glasser, G., Spiess, H.W., Jonas, U.: Site-selective growth of colloidal crystals with photonic properties on chemically patterned surfaces. Adv. Mater. **15**, 1025–1028 (2003)
28. Míguez, H., Chomski, E., García-Santamaría, F., Ibisate, M., John, S., López, C., Meseguer, F., Mondia, J.P., Ozin, G.A., Toader, O., van Driel, H.M.: Photonic bandgap engineering in germanium inverse opals by chemical vapor deposition. Adv. Mater. **13**, 1634–1637 (2001)
29. Blanco, A., Chomski, E., Grabtchak, S., Ibisate, M., John, S., Leonard, S.W., López, C., Meseguer, F., Míguez, H., Mondia, J.P., Ozin, G.A., Toader, O., van Driel, H.M.: Large-scale synthesis of a silicon photonic crystal with a complete threedimensional bandgap near 1.5 micrometres. Nature **405**, 437–440 (2000)
30. Vlasov, Y.A., Bo, X.Z., Sturm, J.C., Norris, D.J.: On-chip natural assembly of silicon photonic bandgap crystals. Nature **414**, 289–293 (2001)
31. Velev, O.D., Tessier, P.M., Lenhoff, A.M., Kaler, E.W.: Materials: a class of porous metallic nanostructures. Nature **401**, 548 (1999)
32. Waterhouse, G.I.N., Waterland, M.R.: Opal and inverse opal photonic crystals: fabrication and characterization. Polyhedron **26**(2), 356–368 (2007)
33. Holland, B.T., Blanford, C.F., Stein, A.: Synthesis of macroporous minerals with highly ordered three-dimensional arrays of spheroidal voids. Science **281**, 538–540 (1998)
34. Lei, Z., Li, J., Zhang, Y., Lu, S.: Fabrication and characterization of highly-ordered periodic macroporous barium titanate by the sol–gel method. J. Mater. Chem. **10**, 2629–2631 (2000)
35. Braun, P.V., Wiltzius, P.: Microporous materials: electrochemically grown photonic crystals. Nature **402**, 603–604 (1999)
36. Xu, L., Wiley, J.B., Zhou, W.L., Frommen, C., Malkinski, L., Wang, J.Q., Baughman, R.H., Zakhidov, A.A.: Electrodeposited nickel and gold nanoscale metal meshes with potentially interesting photonic properties. Chem. Comm. **17**, 997–998 (2000)
37. Pérez, N., Hüls, A., Puente, D., González-Viñas, W., Castaño, E., Olaizola, S.M.: Fabrication and characterization of silver inverse opals. Sens. Actuators B **126**, 86–90 (2007)

38. Bartlett, P.N., Birkin, P.R., Ghanem, M.A.: Electrochemical deposition of macroporous platinum, palladium and cobalt films using polystyrene latex sphere templates. Chem. Commun. **17**, 1671–1672 (2000)
39. Subramania, G., Constant, K., Biswas, R., Sigalas, M.M., Ho, K.M.: Optical photonic crystals synthesized from colloidal systems of polystyrene spheres and nanocrystalline titania. J. Lightwave Technol. **17**, 1970–1974 (1999)
40. Chung, Y., Leu, I., Lee, J., Hona, M.: Fabrication and characterization of photonic crystals from colloidal processes. J. Cryst. Growth **275**, e2389–e2394 (2005)
41. Kulinowski, K.M., Jiang, P., Vaswani, H., Colvin, V.L.: Porous metals from colloidal templates. Adv. Mater. **12**, 833–838 (2000)
42. Velev, O., Lenhoff, A.: Colloidal crystals as templates for porous materials. Curr. Opin. Colloid Interface Sci. **5**, 56–63 (2000)
43. Braun, P.V., Wiltzius, P.: Macroporous materials-electrochemically grown photonic crystals. Curr. Opin. Colloid Interface Sci. **7**, 116–123 (2002)
44. Kuai, S.L., Bader, G., Hache, A., Truong, V.V., Hu, X.F.: High quality ordered macroporous titania films with large filling fraction. Thin Solid Films **483**, 136–139 (2005)
45. Yu, X., Lee, Y.J., Furstenberg, R., White, J.O., Braun, P.V.: Filling fraction dependent properties of inverse opal metallic photonic crystals. Adv. Mater. **19**, 1689–1692 (2007)
46. Subramanian, G., Manoharan, V.N., Thorne, J.D., Pine, D.J.: Ordered macroporous materials by colloidal assembly: a possible route to photonic bandgap materials. Adv. Mater. **11**, 1261–1265 (1999)
47. Wang, B., Zhao, W., Chen, A., Chua, S.J.: Formation of nanoimprinting mould through use of nanosphere lithography. J. Cryst. Growth **288**, 200–204 (2006)
48. Lipson, A.L., Comstock, D.J., Hersam, M.C.: Nanoporous templates and membranes formed by nanosphere lithography and aluminum anodization. Small **5**(24), 2807–2811 (2009)
49. Deckman, H.W., Dunsmuir, J.H.: Natural lithography. Appl. Phys. Lett. **41**, 377–379 (1982)
50. Li, W., Zhao, W., Sun, P.: Fabrication of highly ordered metallic arrays and silicon pillars with controllable size using nanosphere lithography. Phys. E **41**, 1600–1603 (2009)
51. Zhang, Y., Wang, X., Wang, Y., Liu, H., Yang, J.: Ordered nanostructures array fabricated by nanosphere lithography. J. Alloys Compd. **452**, 473–477 (2008)
52. Haynes, C.L., McFarland, A.D., Smith, M.T., Hulteen, J.C., Van Duyne, R.P.: Angle-resolved nanosphere lithography: manipulation of nanoparticle size, shape, and interparticle spacing. J. Phys. Chem. B **106**, 1898–1902 (2002)
53. Li, W., Xu, L., Zhao, W.M., Sun, P., Huang, X.F., Chen, K.J.: Fabrication of large-scale periodic silicon nanopillar arrays for 2D nanomold using modified nanosphere lithography. Appl. Surface Sci. **253**, 9035–9038 (2007)
54. Jeong, G.H., Park, J.K., Lee, K.K., Jang, J.H., Lee, C.H., Kang, H.B., Yang, C.W., Suh, S.J.: Fabrication of low-cost mold and nanoimprint lithography using polystyrene nanosphere. Microelectron. Eng. **87**, 51–55 (2006)
55. Canpean, V., Astilean, S.: Extending nanosphere lithography for the fabrication of periodic arrays of subwavelength metallic nanoholes. Mater. Lett. **63**, 2520–2522 (2009)
56. Patrini, M., Galli, M., Belotti, M., Andreani, L.C., Guizzetti, G., Pucker, G., Lui, A., Bellutti, P., Pavesi, L.: Optical response of one-dimensional (Si/SiO_2)m photonic crystals. J. Appl. Phys. **92**, 1816–1820 (2009)
57. Yablonovitch, E., Gmitter, T.J., Leung, K.M.: Photonic band structure: the face-centered-cubic case employing nonspherical atoms. Phys. Rev. Lett. **67**, 2295–2298 (1991)
58. Özbay, E., Abeyta, A., Tuttle, G., Tringides, M., Biswas, R., Chan, C.T., Soukoulis, C.M., Ho, K.M.: Measurement of a three-dimensional photonic band gap in a crystal structure made of dielectric rods. Phys. Rev. B **50**, 1945–1948 (1994)
59. Qi, M., Lidorikis, E., Rakich, P.T., Johnson, S.G., Joannopoulos, J.D., Ippen, E.P., Smith, H.I.: A three-dimensional optical photonic crystal with designed point defects. Nature **429**, 538–542 (2004)

60. Kuo, C.Y., Lu, S.Y.: Opaline metallic photonic crystals possessing complete photonic band gaps in optical regime. Appl. Phys. Lett. **92**, 121919–121921 (2008)
61. Eradat, N., Huang, J.D., Vardeny, Z.V., Zakhidov, A.A., Khayrullin, I., Udod, I., Baughman, R.H.: Optical studies of metal-infiltrated opal photonic crystals. Synth. Metals **116**, 501–504 (2001)
62. Allard, M., Sargent, E.H., Kumacheva, E., Kalinina, O.: Characterization of internal order of colloidal crystals by optical diffraction. Opt. Q. Electron. **34**, 27–36 (2002)
63. López, C., Vázquez, L., Meseguer, F., Mayoral, R., Ocaña, M., Míguez, H.: Photonic crystal made by close packing SiO_2 submicron spheres. Superlattices Microstruct. **22**, 399–404 (1997)
64. Mizeikis, V., Juodkazis, S., Marcinkevicius, A., Matsuo, S., Misawa, H.: Tailoring and characterization of photonic crystals. J. Photochem. Photobiol. C **2**, 35–69 (2001)
65. Vlasov, Y.A., Deutsch, M., Norris, D.J.: Single-domain spectroscopy of self-assembled photonic crystals. Appl. Phys. Lett. **76**, 1627–1629 (2000)
66. Luo, D.H., Levy, R.A., Hor, Y.F., Federici, J.F., Pafchek, R.M.: An integrated photonic sensor for in situ monitoring of hazardous organics. Sens. Actuators B Chem. **92**, 121–126 (2003)
67. Block, I.D., Chan, L.L., Cunningham, B.T.: Photonic crystal optical biosensor incorporating structured low-index porous dielectric. Sens. Actuators B Chem. **120**, 187–193 (2006)
68. Alexeev, V.L., Das, S., Finegold, D.N., Asher, S.A.: Photonic crystal glucose-sensing material for noninvasive monitoring of glucose in tear fluid. Clin. Chem. **50**, 2353–2360 (2004)
69. Johnson, S.G., Villeneuve, R.L., Fan, S., Joannopoulos, J.D.: Linear waveguides in photonic-crystal slabs. Phys. Rev. B **62**, 8212–8222 (2000)
70. Assefa, S., Petrich, G.S., Kolodziejski, L.A., Mondol, M.K., Smith, H.I.: Fabrication of photonic crystal waveguides composed of a square lattice of dielectric rods. J. Vac. Sci. Technol. B **22**, 3363–3365 (2004)
71. O'Faolain, L., Yuan, X., McIntyre, D., Thoms, S., Chong, H., De La Rue, R.M., Krauss, T.F.: Low-loss propagation in photonic crystal waveguides. Electron. Lett. **42**, 1454–1455 (2006)
72. Joannopoulos, J.D., Johnson, S.G., Winn, J.N., Meade, R.D.: Photonic Crystals: Molding the Flow of Light, 2nd edn. Princeton University Press, Princeton (2008)
73. Miyai, E., Sakai, K., Okano, T., Kunishi, W., Ohnishi, D., Noda, S.: Photonics: lasers producing tailored beams. Nature **441**, 946 (2006)
74. Ritchie, R.H.: Plasma losses by fast electrons in thin films. Phys. Rev. **106**, 874–881 (1957)
75. Barnes, W.L., Dereux, A., Ebbesen, T.W.: Surface plasmon subwavelength optics. Nature **424**, 824–830 (2003)
76. Faraday, M.: Experimental relations of gold (and other metals) to light. Philos. Trans. R. Soc. Lond. **147**, 145–181 (1857)
77. Ritchie, R.H., Arakawa, E.T., Cowan, J.J., Hamm, R.N.: Surface-plasmon resonance effect in grating diffraction. Phys. Rev. Lett. **21**, 1530–1533 (1968)
78. Ebbesen, T.W., Lezec, H.J., Ghaemi, H.F., Thio, T., Wolff, P.A.: Extraordinary optical transmission through sub-wavelength hole arrays. Nature **391**, 667–669 (1998)
79. Martín-Moreno, L., García-Vidal, F.J., Lezec, H.J., Pellerin, K.M., Thio, T., Pendry, J.B., Ebbesen, T.W.: Theory of extraordinary optical transmission through subwavelength hole arrays. Phys. Rev. Lett. **86**, 1114–1117 (2001)
80. Sauvan, C., Billaudeau, C., Collin, S., Bardou, N., Pardo, F., Pelouard, J.L., Lalanne, P.: Surface plasmon coupling on metallic film perforated by twodimensional rectangular hole array. Appl. Phys. Lett. **92**, 011125–011127 (2008)
81. Billaud, P., Huntzinger, J.R., Cottancin, E., Lermé, J., Pellarin, M., Arnaud, L., Broyer, M., Del Fatti, N., Vallée, F.: Optical extinction spectroscopy of single silver nanoparticles. Eur. Phys. J. D **43**, 271–274 (2007)

82. Chan, G.H., Zhao, J., Schatz, G.C., Van Duyne, R.P.: Localized surface plasmon resonance spectroscopy of triangular aluminum nanoparticles. J. Phys. Chem. C **112**, 13958–13963 (2008)
83. Hicks, E.M., Lyandres, O., Hall, W.P., Zou, S., Glucksberg, M.R., Van Duyne, R.P.: Plasmonic properties of anchored nanoparticles fabricated by reactive ion etching and nanosphere lithography. J. Phys. Chem. C **111**, 4116–4124 (2007)
84. Haynes, C.L., Van Duyne, R.P.: Nanosphere lithography: a versatile nanofabrication tool for studies of size-dependent nanoparticle optics. J. Phys. Chem. B **105**, 5599–5611 (2001)
85. Haes, A.J., Chang, L., Klein, W.L., Van Duyne, R.P.: Detection of a biomarker for Alzheimer's Disease from synthetic and clinical samples using a nanoscale optical biosensor. J. Am. Chem. Soc. **127**, 2264–2271 (2005)
86. Li, H., Luo, X., Du, C., Chen, X., Fu, Y.: Ag dots array fabricated using laser interference technique for biosensing. Sens. Actuators B, **134**, 940–944 (2008)
87. Bingham, J.M., Willets, K.A., Shah, N.C., Andrews, D.Q., Van Duyne, R.P.: Localized surface plasmon resonance imaging: simultaneous single nanoparticle spectroscopy and diffusional dynamics. J. Phys. Chem. C **113**, 16839–16842 (2009)
88. McFarland, A.D., Van Duyne, R.P.: Single silver nanoparticles as real-time optical sensors with zeptomole sensitivity. Nano Lett. **3**, 1057–1062 (2003)
89. Yonzon, C.R., Stuart, D.A., Zhang, X., McFarland, A.D., Haynes, C.L., Van Duyne, R.P.: Towards advanced chemical and biological nanosensors—an overview. Talanta **67**, 438–448 (2005)
90. Homola, J., Yeea, S.S., Gauglitzb, G.: Surface plasmon resonance sensors: review. Sens. Actuators B Chem. **54**, 3–15 (1999)
91. Hutter, E., Fendler, J.H.: Exploitation of localized surface plasmon resonance. Adv. Mater. **16**, 1685–1706 (2004)
92. Liedberg, B., Nylander, C., Lundstrom, I.: Biosensing with surface plasmon resonance—how it all started. Biosens. Bioelectron. **10**, i–ix (1995)
93. Vikinge, T.P., Askendal, A., Liedberg, B., Lindahl, T., Tengvall, P.: Immobilized chicken antibodies improve the detection of serum antigens with surface plasmon resonance (spr). Biosens. Bioelectron. **13**, 1257–1262 (1998)
94. Carlsson, J., Gullstrand, C., Westermark, G.T., Ludvigsson, J., Enander, K., Liedberg, B.: An indirect competitive immunoassay for insulin autoantibodies based on surface plasmon resonance. Biosens. Bioelectron. **24**, 876–881 (2008)
95. Lee, K.L., Wang, W.S., Wei, P.K.: Sensitive label-free biosensors by using gap plasmons in gold nanoslits. Biosens. Bioelectron. **24**, 210–215 (2008)
96. Dostalek, J., Homola, J.: Surface plasmon resonance sensor based on an array of diffraction gratings for highly parallelized observation of biomolecular interactions. Sens. Actuators B Chem. **129**, 303–310 (2008)
97. Nylander, C., Liedberg, B., Lind, T.: Gas detection by means of surface plasmon resonance. Sens. Actuators **3**, 79–88 (1982–1983)
98. Liedberg, B., Nylander, C., Lunstrom, I.: Surface plasmon resonance for gas detection and biosensing. Sens. Actuators **4**, 299–304 (1983)
99. Chadwick, B., Tann, J., Brungs, M., Gal, M.: A hydrogen sensor based on the optical generation of surface plasmons in a palladium alloy. Sens. Actuators B Chem. **17**, 215–220 (1994)
100. Arakawa, T., Miwa, S.: Selective gas detection by means of surface plasmon resonance sensors. Thin Solid Films **281–282**, 466–468 (1996)
101. Manera, M., Spadavecchia, J., Buso, D., de Julian Fernandez, C., Mattei, G., Martucci, A., Mulvaney, P., Perez-Juste, J., Rella, R., Vasanelli, L., Mazzoldi, P.: Optical gas sensing of TiO_2 and TiO_2/Au nanocomposite thin films. Sens. Actuators B Chem. **132**, 107–115 (2008)
102. de Julian Fernandez, C., Manera, M., Pellegrini, G., Bersani, M., Mattei, G., Rella, R., Vasanelli, L., Mazzoldi, P.: Surface plasmon resonance optical gas sensing of nanostructured ZnO films. Sens. Actuators B Chem. **130**, 531–537 (2008)

103. Özbay, E.: Plasmonics: merging photonics and electronics at nanoscale dimensions. Science **311**, 189–193 (2006)
104. Atwater, H., Maier, S., Polman, A., Dionne, J., Sweatlock, L.: The new "p–n junction": plasmonics enables photonic access to the nanoworld. MRS Bull. **30**, 385–389 (2005)
105. Hobson, P., Wedge, S., Wasey, J., Sage, I., Barnes, W.: Surface plasmon mediated emission from organic light-emitting diodes. Adv. Mater. **14**, 1393–1396 (2002)
106. Okamoto, K., Niki, I., Shvartser, A., Narukawa, Y., Mukai, T., Scherer, A.: Surface plasmon-enhanced light emitters based on InGaN quantum wells. Nat. Mater. **3**, 601–605 (2004)
107. Westphalen, M., Kreibig, U., Rostalski, J., Luth, H., Meissner, D.: Metal cluster enhanced organic solar cells. Sol. Energy Mater. Sol. Cells **61**, 97–105 (2000)
108. Tredicucci, A., Gmachl, C., Capasso, F., Hutchinson, A.L., Sivco, D.L., Cho, A.Y.: Single-mode surface-plasmon laser. Appl. Phys. Lett. **76**, 2164–2166 (2000)
109. Campion, A., Kambhampati, P.: Surface-enhanced Raman scattering. Chem. Soc. Rev. **27**, 241–250 (1998)

Ion Beam Sputtering: A Route for Fabrication of Highly Ordered Nanopatterns

Marina Cornejo, Jens Völlner, Bashkim Ziberi, Frank Frost and Bernd Rauschenbach

Abstract This chapter focuses on the self-organized pattern formation by ion beam sputtering. A general description and experimental observations are presented, showing the complexity of the processes involved but also its great potential as patterning technique. The main focus is set on the pattern formation on silicon surfaces. It is shown that several experimental parameters are involved in the topography evolution. Namely, the influence of the ion incidence angle, ion energy, fluence, sample manipulation and substrate temperature is discussed. Additionally, evidence of the importance of iron incorporation in the formation of certain features is presented. The possibility of applying this technique to other materials is illustrated with examples on germanium, compounds semiconductor, silica and crystalline metals.

1 Introduction

Ion beam sputtering, i.e. the removal of atoms from a solid substrate due to the impingement of energetic charged particles, is used in several processing and analytical techniques. Due to the interaction of the ions with the substrate, the surface topography is modified and under certain conditions regular nanostructures can evolve by self-organization processes. The diversity of nanostructures that can be formed in only one step in a wide variety of materials (e.g. elemental and compound semiconductors, mono and polycrystalline metals, dielectrics) makes

---

M. Cornejo (✉), J. Völlner, B. Ziberi, F. Frost and B. Rauschenbach
Leibniz-Institut für Oberflächenmodifizierung e.V., Permoserstr. 15,
04318 Leipzig, Germany
e-mail: marina.cornejo@iom-leipzig.de

F. A. Lasagni and A. F. Lasagni (eds.), *Fabrication and Characterization in the Micro-Nano Range*, Advanced Structured Materials, 10,
DOI: 10.1007/978-3-642-17782-8_4, © Springer-Verlag Berlin Heidelberg 2011

this technique an attractive alternative route for the production of nanopatterned surfaces.

Navez observed in 1962 that wave-like structures (ripples) evolved on glass surfaces after the irradiation with air ions [1]. The direction of the ripples was observed to be dependent on the incidence angle. For ion beam incidence close to the normal the ripples were oriented perpendicular to the projection of the ion beam on the surface (wave vector parallel to the ion beam projection), and they were parallel (wave vector perpendicular) when the incidence angle was close to grazing angles.

Since then this phenomenon has been intensively studied and a certain control on the self-organized nanostructures has been achieved. However, there are still many issues to be solved from the theoretical and experimental point of view. There are several process parameters that control the topography evolution: ion species, ion energy, incidence angle, fluence, flux, substrate temperature, among others. Latest studies reported also growing evidence of the importance of co-deposition of metals during the irradiation. The many operational parameters that determine the topography evolution indicate that there are many degrees of freedom for tailoring the surface topography but also make the understanding of the phenomenon a challenge.

There is a wide variety of nanostructures that can be fabricated by ion induced self-organization on different materials. Some of the different topographies formed on Si and Ge under different experimental conditions are shown in Fig. 1. Two interesting types of patterns are dots and wave-like features (here called ripples), which can be quite regular under certain conditions.

The physical mechanisms behind the morphological evolution of the surface have been intensively investigated. Typically, the surface of the substrate is far

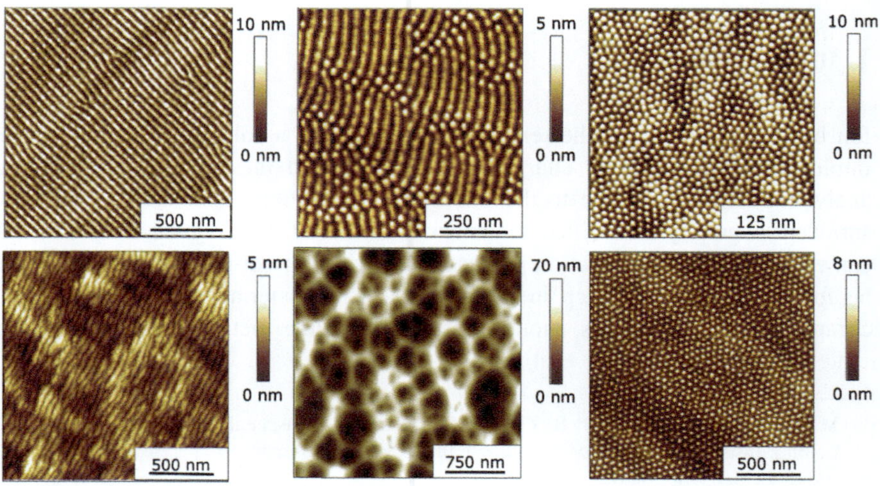

Fig. 1 AFM images of different topographies produced on Si and Ge surfaces by low-energy ion beam erosion under different experimental conditions

from equilibrium during the sputtering and many atomistic surface processes become effective. There are many theoretical models that try to explain the resulting topography. However, a complete understanding of the physical processes has not been achieved yet. Most models are based on Bradley and Harper (BH) model [2].

This linear approach can explain some experimental observations at short times. More generalized theories that consider the stochastic nature of the ion arrival to the substrate, include high-order linear and nonlinear effects and consider additional relaxation mechanisms [3–9]. According to the widely accepted BH model, based on Sigmund's theory [10], the nanostructures are formed as a consequence of the interplay between roughening due to the curvature dependent sputtering and the smoothing due to surface diffusion processes.

In this chapter, on rather a descriptive level, the possibilities offered by the self-organization by ion beam sputtering will be presented. It will be shown that a certain control of the topography evolution can be achieved by choosing the appropriate experimental conditions. In Sect. 2 the variety of experimental conditions at which the self-organization takes place will be given, i.e. ion species, target materials, ion sources, etc. In Sect. 3 the focus will be set on the patterning on silicon and the effect of different experimental parameters. Additionally, and in order to address an important issue which is being intensively investigated at the moment, the role of metal incorporation will be also discussed. In Sect. 4 some examples for other materials will be presented.

2 Experimental Conditions

For the self-organized patterning by ion beam erosion ions from inert gases are usually used. The differences in the self-organized nanostructures formed on Si and Ge using Ar^+, Kr^+ and Xe^+ are in some cases not significant and not easily correlated with the difference in atomic number and atomic weight of the ions. However, there is a limitation with respect to the relation of the ion and substrate masses. No structure seems to evolve when the ion is lighter than the substrate atoms. This seems to be the reason why the pattern formation on Ge surfaces is not achieved using Ar and Ne is not appropriate for the patterning of Si [11]. The explanation for this effect may be related with the distribution of the deposited energy. Using heavier ions, the energy distribution is concentrated closer to the surface and, in consequence, more recoils are produced. Another reason could be related to the fact that for decreasing ion mass the ejected atoms from the substrate and backscattered projectile ions become more important and both contribute to the preferential erosion of peaks compared with valleys, thus leading to additional smoothening.

According to the published studies focused and unfocused ion beams can be used and with varying diameters. The use of broad beam sources (ion beam diameter from 3 to 20 cm or larger) represents an advantage for potential industrial applications. The sources used can vary not only in the size of the beam produced but also in the way the ions are created. A widely used ion source is the so-called Kaufman source, where a hot filament is heated by AC or DC to generate the electron emission to ionize the working gas [12]. In the ECR ion sources, on the other hand, microwave energy is coupled to the ion generating discharge through electron cyclotron resonance [13]. In other types of ion sources the ions can be generated, for example, by radio frequency, electron beam and laser.

Even when low energies (up to 2 keV) are mostly used [14–24], the self-organization has also been observed at higher energies (4–50 keV) [1, 25–27].

The self-organized pattern formation by ion beam sputtering can be applied to a wide variety of materials. It has been studied on single-crystalline semiconductors (Si, Ge, compound semiconductors) [15–18, 24, 26, 28–30], single-crystalline metals (Cu, Ag) [31–34], polycrystalline metals (Ag, Au, Pt) [35–37] and amorphous materials (SiO_2) [19, 38–40].

The resulting topography is highly affected by the experimental conditions: angle of incidence of the ions, fluence, substrate temperature, ion energy, flux, divergence of the ions within the beam, etc. Sample manipulation during the irradiation also affects the topography evolution; i.e. simultaneous sample rotation. Due to the several parameters that control the topography evolution and the recently discovered potential importance of metal incorporation it is difficult to compare the different published experimental results making the experimental and theoretical study of the technique very complex.

3 Self-organized Patterning on Si

Due to its technological importance and the simplicity provided as a one-component material Si has been intensively studied. The different topographies formed by the ion induced self-organization processes on Si and their dependence on the experimental parameters have been investigated in the last decades. Another reason to study Si is that after some minutes of irradiation, even at low ion energies, a surface layer of the crystalline material amorphisizes eliminating the effects that could be related to the crystalline structure of the material. This is valid, in general, for semiconductor materials at temperatures up to some hundreds degree Celsius. Here the effect of some experimental parameters in the pattern formation on Si will be presented.

3.1 Ion Incidence Angle

It is known since the self-organization by ion beam sputtering was first observed that the incidence angle α (the angle formed by the ion beam and the substrate normal) plays an important role on the topography evolution [1, 2, 26]. In Fig. 2

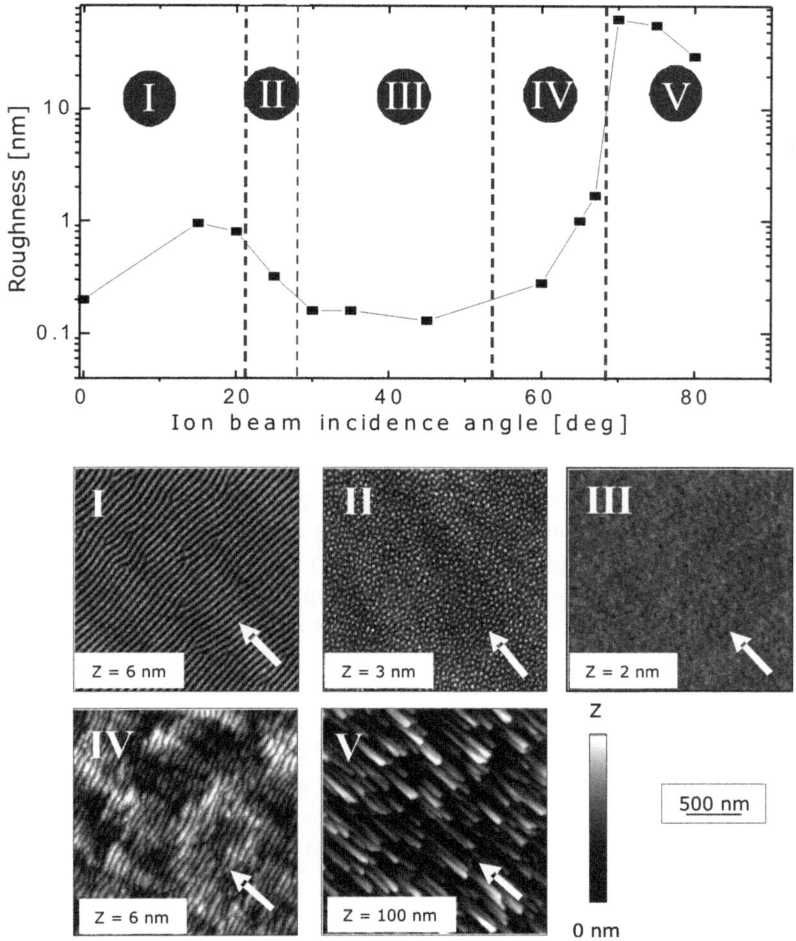

Fig. 2 Surface roughness vs. ion beam incidence angle for Si(001) bombarded with Kr^+ ions, $E_{ion} = 2$ keV, $U_{acc} = -1$ kV, $\Phi = 6.7 \times 10^{18}$ cm^{-2}, without sample rotation. AFM images of the different topographies (examples): I: ripples at near normal incidence, II: dots, III: smooth surface, IV: ripples at high incidence angles and V: columnar structures. The *white arrows* indicate the projection of the ion beam on the surface

the different topographies formed on Si(001) by irradiation at room temperature with Kr^+ with ion energy E_{ion} of 2 keV at different incidence angles without simultaneous sample rotation using a Kaufman type source are presented. The ion current density j_{ion} was 300 μA cm^{-2} (corresponding to an ion flux J of 1.87×10^{15} cm^{-2} s^{-1}) and the total fluence Φ 6.7×10^{18} cm^{-2}. For broad beam ion sources (e.g. Kaufman type) there is an additional experimental parameter that seems to be relevant for the processing: the acceleration voltage U_{acc}. These sources posses an extraction system complied by two or three grids, which are at different voltages. The source used here has two grids, the screen and accelerator grid. The potential applied at the first grid determines the ion energy and the acceleration voltage applied at the second one determines the divergence and angular distribution of the ions within the beam. U_{acc} can vary from -0.2 to -1 kV. $U_{acc} = -1$ kV, which was used for the results presented in Fig. 2, corresponds to the highest divergence. More details about the broad beam ion source used are given elsewhere [41]. As it can be observed in Fig. 2, at near normal incidence, i.e. α from 5 to 25°, ripples are formed (Fig. 2 I). These ripples show a relatively high regularity and choosing the appropriate operating parameters the wavelength and the amplitude can be tuned between ~30–70 nm and up to ~8 nm, respectively. At an incidence angle near 45° the surface smoothens due to the sputtering (Fig. 2 III). The smoothing between α from ~35° to 60° is nearly independent of E_{ion} and ion species used and it has been also intensively studied [42, 43]. It represents an important application of this technique. The dot-like structures shown in Fig. 2 II only evolve in a narrow window of operating conditions and in some cases appear as a mixture of dots and ripples. At higher angles ripples and columnar structures are formed (Fig. 2 IV and V).

The direction of the anisotropic nanostructures is determined by the direction of the ion beam. The ripples (Fig. 2 I and IV) are perpendicular to the projection of the beam on the surface while the columnar structures formed at grazing angles (Fig. 2 V) are parallel. The ripples at higher angles and the columnar structures (Fig. 2 IV and V) were already reported in the first published studies about this phenomenon even under different irradiation conditions [1, 26]. Carter et al. reported in 1977 the formation of perpendicular-mode ripples at $\alpha = 45°$ and parallel columnar structures at $\alpha = 75°$ on Si by sputtering with 40 keV Ar^+. This change in the structure orientation was successfully predicted by Bradley and Harper model [2]. There are many reports about the ripples and columnar structures on Si obtained at large angles [16, 17, 24, 27, 44, 45].

In contrast to that predicted by BH model and other theories, the transition between the different types of nanostructures is continuous. The transition from the perpendicular-mode ripples to the parallel columnar structure on Si(001) is presented in Fig. 3. Perpendicular-mode ripples are observed at $\alpha = 65°$ (Fig. 3a). Increasing the incidence angle high amplitude structures parallel to the ion beam direction emerge and they dominate the topography at high angles (Fig. 3f). In Fig. 4 SEM images of Si(001) after bombardment at $\alpha = 70°$ (Fig. 4a), and $\alpha = 75°$ (Fig. 4b) are shown. At $\alpha = 75°$, the columnar structures parallel to the ion beam are the only features observed, however at $\alpha = 70°$, the columnar structures and the perpendicular-mode ripples coexist.

The results shown in Fig. 2 are valid for an ion energy range of 1–2 keV for the experimental conditions described above and for Ar^+, Kr^+, and Xe^+. However, there are discrepancies in the literature about the incidence angle effect. Ozaydin-Ince et al. [46] reported the formation of ripples on Si after the irradiation with Ar^+ 0.5 and 1 keV at $\alpha = 45–60°$ and not smooth surface. Madi et al. [47] observed the smoothening of the Si surface during the irradiation with Ar^+ 1 keV at an incident angle of 10° instead of ripple formation. According to Zhang et al. [27] no pattern

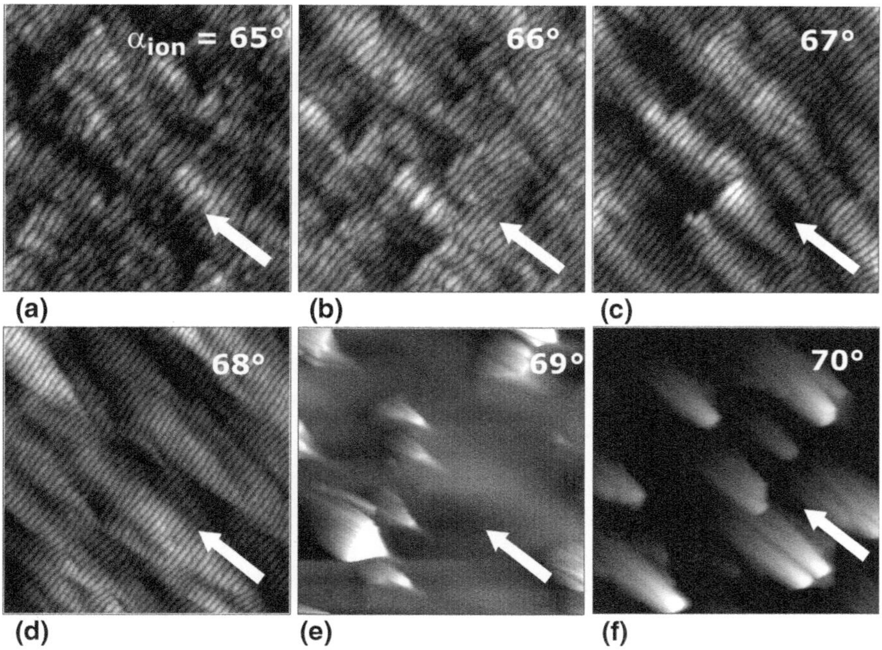

Fig. 3 AFM images of Si(001) surfaces after irradiation with Kr^+, $E_{ion} = 2$ keV, $U_{acc} = -1$ kV, $\Phi = 6.7 \times 10^{18}$ cm^{-2}, at $\alpha = 65$–$70°$. All the images are 2 × 2 μm with a vertical scale of **a** 8 nm, **b**, **c** 10 nm, **d** 13 nm, **e** 170 nm, and **f** 250 nm

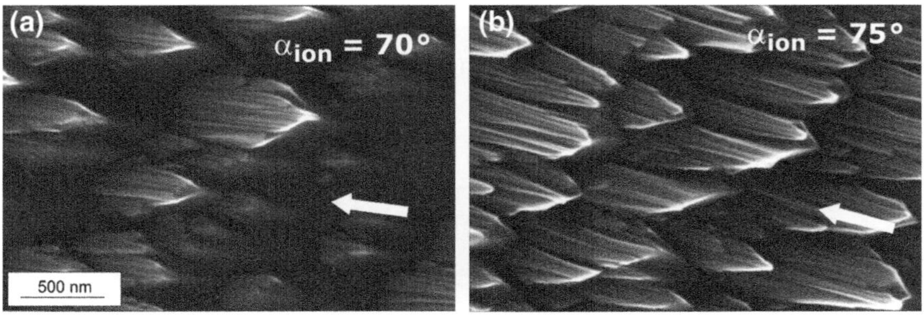

Fig. 4 SEM images (*top view*) of Si(001) surfaces after irradiation with Kr^+, $E_{ion} = 2$ keV, $U_{acc} = -1$ kV, $\Phi = 6.7 \times 10^{18}$ cm^{-2}, at $\alpha =$ **a** 70° and **b** 75°

evolves on Si bombarded with Xe^+ at $\alpha = 30°$ with different ion energy (from 1 to 50 keV). These are only some examples that show the disagreement on the experimental results reported.

There are also disparities in the observations for bombardment of Si with Ar^+ at normal incidence. On Si(001) sputtered with Ar^+ at normal incidence the formation of holes without regular distribution was observed at $E_{ion} = 0.5$ keV by

Ziberi et al. [48]. These observations are consistent with the report from Madi et al. [49], who reported the formation of holes with $E_{ion} < 600$ eV. However, while Ozaydin et al. [50] observed no pattern evolution by sputtering of Si(001) with Ar^+ $E_{ion} = 1$ keV, under the same experimental conditions Gago et al. [51] reported the formation of nanodots. The formation of dots was also observed with $E_{ion} = 1.2$ keV [15] and $E_{ion} = 1.8$ keV [48].

The formation of perpendicular-mode ripples on Si at $\alpha = 30°$ was reported by Habenicht et al. [52] but under very different conditions; they produced ripples using focused ion beam, 30 keV Ga^+. The ripples formed by sputtering with noble gas ions at near normal incidence have been presented on Si and Ge only by our group [11, 20, 53] and on Ge by Carbone et al. [18]. A possible explanation for this fact and also for the differences in the topography evolution at normal incidence may be related to the inadvertent incorporation of metallic atoms during sputtering. This issue will be addressed in Sect. 3.4.

3.2 Sample Rotation

The production of arrays of nanodots on Si surfaces results very attractive for many technological applications. In the Sect. 3.1 the topography evolution of Si under ion beam sputtering at different incident angles without rotation of the sample was discussed. The dots can be formed on Si surfaces at normal incidence [15, 48] or at α near 30° choosing the appropriate operational parameters (see Fig. 2 II). Additionally, it has been found that domains of hexagonally ordered dots formed on Si irradiated at higher ion incidence angles (e.g. $\alpha = 75°$) with simultaneous sample rotation. These dots have a high degree of ordering. Fig. 5 shows the resulting topography on Si(001) after irradiation with Kr^+,

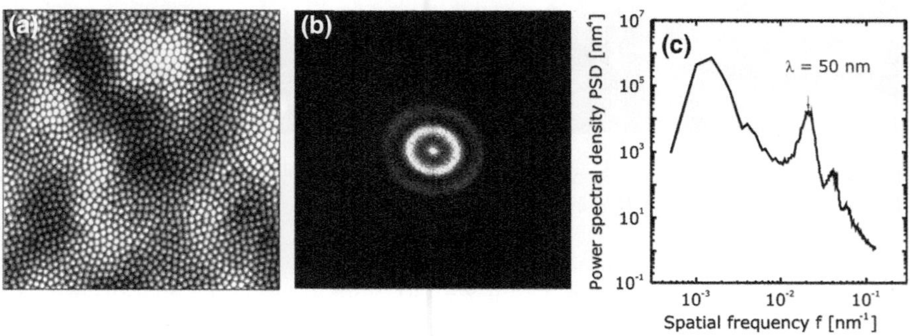

Fig. 5 Si(001) after bombardment with Kr^+, $E_{ion} = 2$ keV, $U_{acc} = -1$ kV, $\Phi = 6.7 \times 10^{18}$ cm^{-2}, $\alpha = 75°$ with simultaneous rotation of the sample (12 rotations per minute). **a** AFM image, 2 × 2 μm and vertical scale of 30 nm. **b** Fast Fourier transformation image (FFT) with a frequency scale of −128 to 128 μm^{-1}. **c** Power spectral density diagram (PSD)

$E_{ion} = 2$ keV, $U_{acc} = -1$ kV, $\alpha = 75°$, $\Phi = 6.7 \times 10^{18}$ cm^{-2}, with simultaneous rotation (12 rotations/min). The ring-like Fourier spectrum (Fig. 5b) shows the random azimuthal distribution of the individual domains, and the adjoining rings the narrow size distribution of the dots. The first peak in the PSD diagram (Fig. 5c) indicates a mean dot distance of ~50 nm.

3.3 Fluence

According to most experimental studies, the amplitude of perpendicular-mode ripples and dots on Si at room temperature increases with the time up to saturation [20, 24, 48, 54]. In Fig. 6 the temporal evolution of the roughness, which is proportional to the amplitude of the ripples, on Si(001) bombarded with Kr^+ at room temperature, with $E_{ion} = 2$ keV and $U_{acc} = -1$ kV is shown. The data correspond to different ion beam incidence angles at which ripples are formed ($\alpha = 20$, 35, 39, and 65°). The roughness (i.e. the amplitude) saturation is clearly observed. The wavelength of the features is, on the other hand, not affected by the fluence. The amplitude saturation is also observed on other materials (Ge, compounds semiconductors, Cu) and it may be related to the nonlinear effects that become effective after a certain sputtering time. As the nonlinear effects are not contemplated in BH model, this model does not predict saturation.

In the case of Si, after saturation for increasing fluence the size of the ripples is constant but the order increases. This is illustrated in Fig. 7, where AFM images and power spectral density (PSD) diagrams for Si(001) irradiated with Kr^+ at room temperature, with $E_{ion} = 2$ keV and $U_{acc} = -1$ kV and at $\alpha = 35°$ are presented. By direct observation of the AFM images, it is possible to see that with increasing fluence the nanostructures show higher ordering. This can be also inferred from the PSD diagrams. The peak frequency remains almost constant with the time;

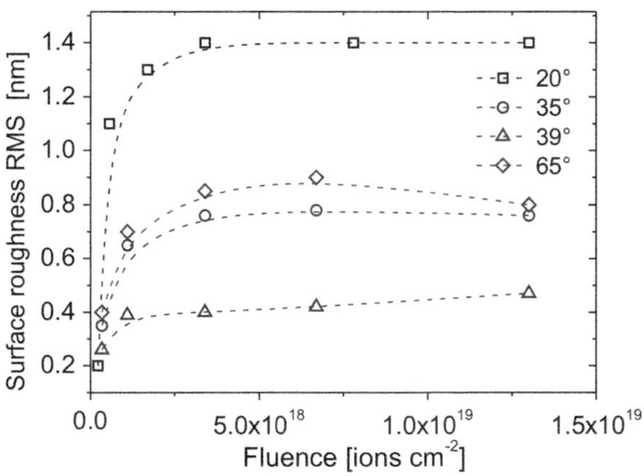

Fig. 6 Temporal evolution of the roughness of Si(001) after irradiation with Kr^+, $E_{ion} = 2$ keV, $U_{acc} = -1$ kV, $\alpha = 20°$, 35°, 39°, and 65°

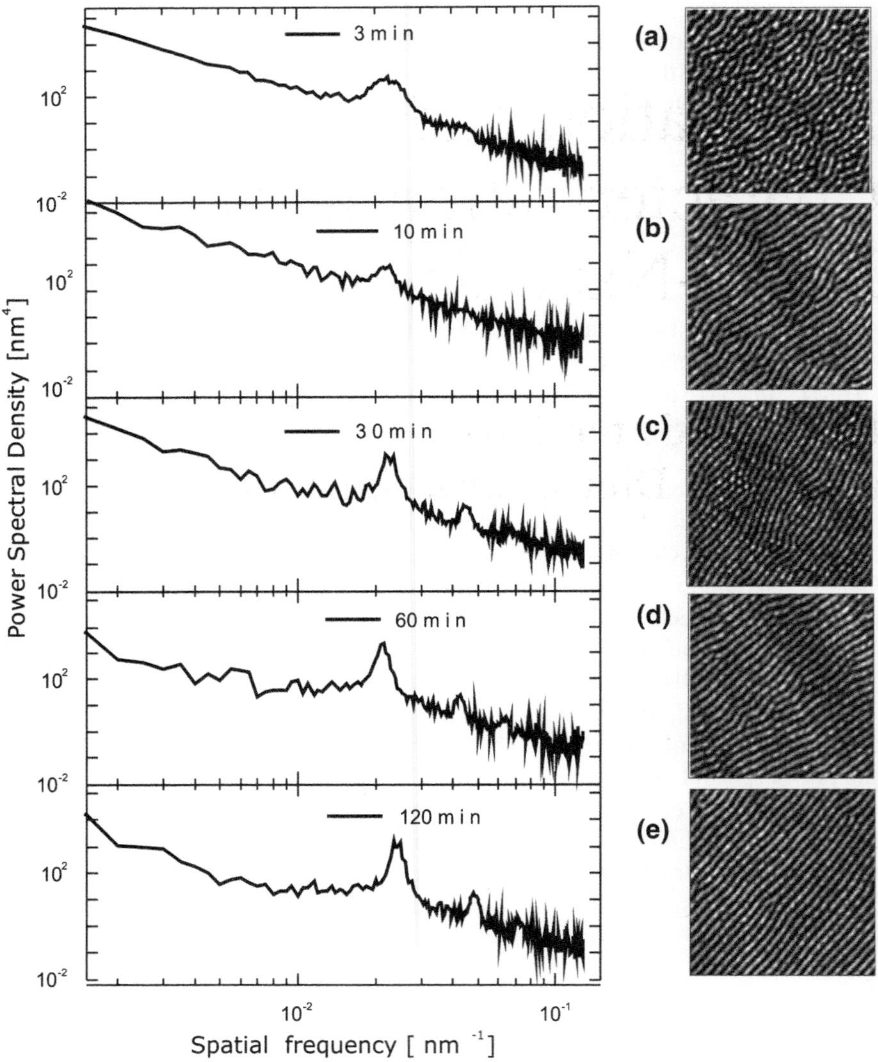

Fig. 7 Temporal evolution of ripples. Power spectral density diagrams and AFM images of Si(001) after irradiation with Kr^+, $E_{ion} = 2$ keV, $U_{acc} = -1$ kV, $\alpha = 35°$, with fluence of **a** 3.4×10^{17} cm^{-2}, **b** 1.1×10^{18} cm^{-2}, **c** 3.4×10^{18} cm^{-2}, **d** 6.7×10^{18} cm^{-2}, and **e** 1.3×10^{19} cm^{-2}. The AFM images are 2×2 μm. The vertical scale is **a** 2.5 nm, **b** 3 nm, **c** 3.5 nm, **d** 4 nm, and **e** 4 nm. The noise observed at high frequencies reflects unfavorable ambient conditions during measurement

i.e. the wavelength of the ripples does not change significantly. Also the narrowing of the peak indicates that the correlation order increases with the fluence. As a measure of the order the correlation length can be used and can be calculated from the full width at half maximum of the first-order peaks of the PSD curves [55].

The temporal evolution of the nanostructures is different at high temperature. Coarsening was observed for parallel- and perpendicular-mode ripples on Si at temperatures between 600 and 750°C ($E_{ion} = 0.25$–1.2 keV, $\alpha = 60°$) [24]. An increase of the characteristic length scale with the fluence was also found at temperatures higher than 400°C on Si(001) with 1 keV Ar^+ normal incidence sputtering [50].

3.4 Ion Energy

The influence of the ion energy also depends on the substrate temperature. At room temperature and using low energies the amplitude and the wavelength or period of the ripples formed at near normal increase with E_{ion}; the wavelength can be tuned from ~35 to 75 nm varying E_{ion} from 0.5 to 2 keV [20]. On the other hand, Brown and Erlebacher [24] observed that at high temperatures (~700°C) as E_{ion} is increased from 0.25 to 1.2 keV (Ar^+), there is a significant decrease in the ripples amplitude and wavelength on Si(111) at $\alpha = 60°$. The observations from Brown and Erlebacher are in accordance with BH model. Using higher ion energy ($E_{ion} = 60$–100 keV), Hazra et al. [56] found that with the increase of E_{ion} the amplitude of the ripples formed on Si(001) at $\alpha = 60°$ decreases while the wavelength increases.

In contrast to ion beam sputtering at high temperature, where the substrate surface remains crystalline, at low temperatures the surface is amorphized. The amorphization begins when the fluence reaches the amorphization threshold, then the thickness of the amorphous layer increases and saturates after a short time. Together with α, E_{ion} has a strong influence in the amorphous layer thickness; with $E_{ion} = 2$ keV the layer is some nanometers deep (~5–8 nm) [11] and with 120 keV it can reach a depth larger than 250 nm [45]. In Fig. 8 a high resolution transmission electron microscopy (HRTEM) cross-section image of Si(001)

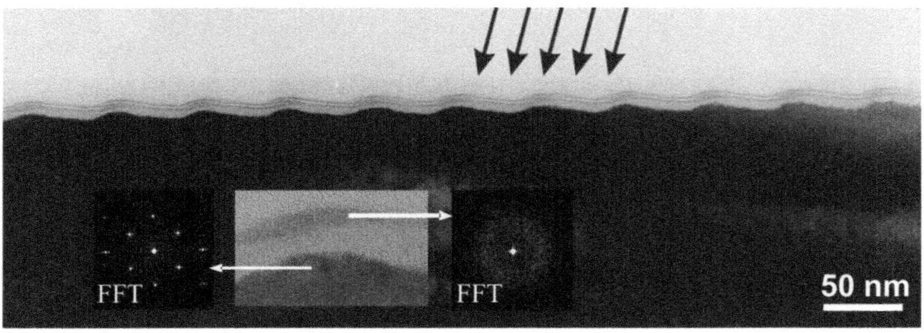

Fig. 8 Cross-sectional HRTEM image of ripple patterns (Si(001), Kr^+, $E_{ion} = 1.2$ keV, $\alpha = 15°$). The *arrows* indicate the direction of the ion beam

irradiated with 1.2 keV Kr^+ at near normal incidence is shown. The ripple wavelength is ~50 nm and the amplitude ~5 nm. The amorphous layer in this case has a depth of ~6 nm. Due to the amorphization of the surface layer, no significant differences have been found in the topography evolution at low temperatures in the different crystal orientations. It can be observed also that the ripples are asymmetric; the side facing the beam is less steep than the other.

The variation of the ion energy affects not only the size of the evolving features but also, under certain conditions, can lead to different type of structures. In Fig. 9 AFM images of the resulting topography on Si(001) irradiated with Kr^+, at $\alpha = 65°$, with a total fluence $\Phi = 6.7 \times 10^{18}\ cm^{-2}$, $U_{acc} = -1$ kV using different E_{ion} are presented. While with $E_{ion} = 2$ keV perpendicular-mode ripples are the only type of feature observed (Fig. 9a), at $E_{ion} = 1$ keV they coexist with waves parallel to the ion beam (Fig. 9b). Decreasing E_{ion} to 0.75 and 0.5 keV high amplitude parallel-mode structures dominate.

Similar effect of the ion energy is presented in Fig. 10. When Si(001) is irradiated with Kr^+, at $\alpha = 20°$, with a total fluence $\Phi = 6.7 \times 10^{18}\ cm^{-2}$, $U_{acc} = -1$ kV, $E_{ion} = 2$ keV (Fig. 10a) perpendicular-mode ripples evolve. However, decreasing E_{ion} some parallel-mode waves appear (Fig. 10c) and at lower E_{ion} perpendicular and parallel-mode ripples are simultaneously present (Fig. 10d) and at $E_{ion} = 0.3$ keV only parallel-mode ripples are observed with AFM (Fig. 10e). In the power spectral density (PSD) diagrams (Fig. 10f) the peaks corresponding to both kinds of ripples are observed. The perpendicular-mode ripples peak (indicated with the open arrows) becomes more defined with increasing E_{ion}, while the parallel-mode ripples peak (full arrows) turns weaker. The lack of definition of the peaks made the quantification of the parallel-mode ripples wavelength more complex. 2D FFT (fast Fourier transformation) diagrams from the amplitude signal from AFM images were used. In Fig 10g the evolution of the wavelength with the ion energy for both types of ripples is shown. The wavelength of the perpendicular-mode ripples increases with E_{ion}. The same effect is observed for the parallel-mode ripples, however, the wavelength is difficult to determine for $E_{ion} > 1$ keV because of the decreasing amplitude.

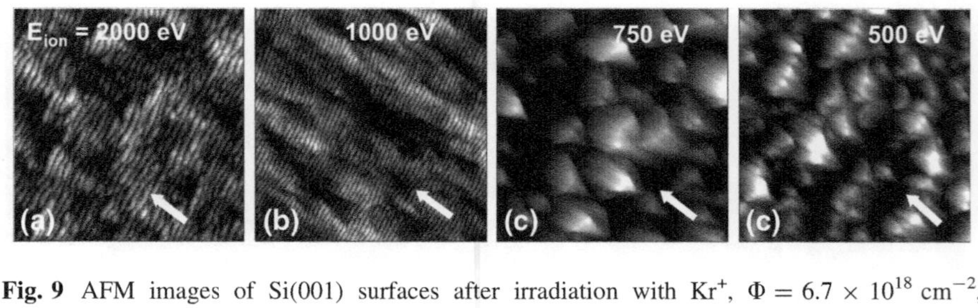

Fig. 9 AFM images of Si(001) surfaces after irradiation with Kr^+, $\Phi = 6.7 \times 10^{18}\ cm^{-2}$, $U_{acc} = -1$ kV, $\alpha = 65°$ and E_{ion} from 0.5 to 2 keV. The images are 2×2 μm with a vertical scale of **a**, **b** 6 nm, **c** 230 nm, and **d** 120 nm

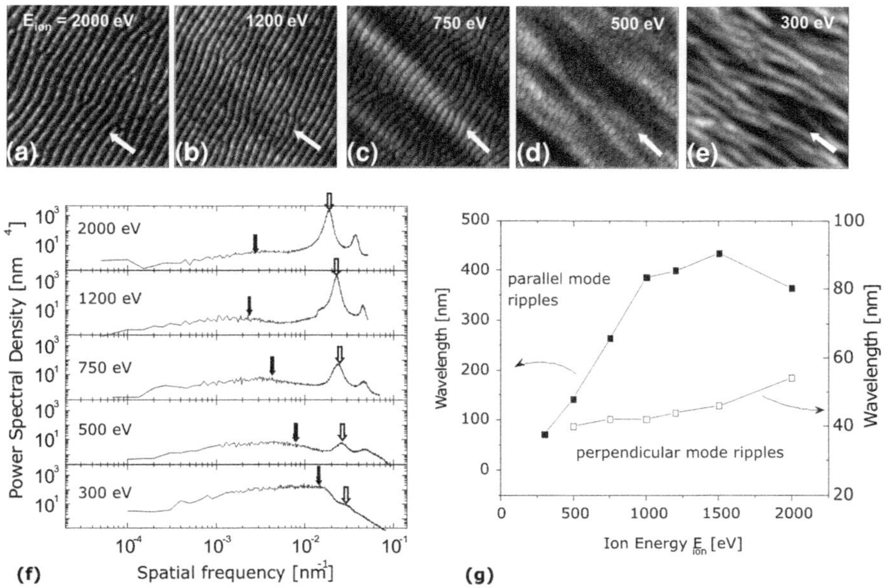

Fig. 10 AFM images (**a–e**) of Si(001) surfaces after irradiation with Kr^+, $\Phi = 6.7 \times 10^{18}$ cm^{-2}, $U_{acc} = -1$ kV, $\alpha = 20°$ and E_{ion} from 0.3 to 2 keV. All the images are 2 × 2 μm with a vertical scale of **a–c** 3 nm, **d** 5 nm, and **e** 12 nm. **f** Power spectral density diagram obtained from the AFM images **a–e**. The *closed* and *open arrows* indicate the position of the peaks of the parallel and perpendicular mode ripples, respectively. **g** Ion energy dependence of the ripple wavelength

3.5 Simultaneous Metal Incorporation

The structuring by ion beam sputtering in the presence of impurities has been first observed several decades ago [57]. Cones or pyramids on the micron size may be formed by ion bombardment with a concurrent supply of a seed material [57–61]. Even when the structures presented in this study are much smaller, it has been suggested that the inadvertent incorporation of impurities may also play an important role in their formation. Ozaydin et al. investigated the relevance of the incorporation of Mo [46, 62, 63] during the ion sputtering of Si surfaces at normal incidence. They found that with the simultaneous incorporation of Mo the formation of nanodots instead of smoothing takes place. The role of the simultaneous metal incorporation was also studied by Sánchez-García et al. [64, 65]. They showed that the sputtering at normal incidence with simultaneous metal incorporation (Fe and Mo) leads to the formation of nanoholes or nanodots on Si surfaces. They observed that the topography could be changed from nanoholes to nanodots by increasing the ion current density or increasing the fluence at low ion current density. They correlated this change from holes to dots with a decrease of the metal content on the substrate surface.

Besides the topography evolution at normal incidence, the incorporation of Fe seems to play also a role in the pattern formation of ripples on Si at near normal incidence presented here (Fig. 2 I). With our experimental setup, described briefly in Sect. 3.1 and more in detail in [41], the incorporation of foreign atoms coming from the sputtering of the chamber components can not be completely avoided. In particular, the sputtering from a stainless steel plate lining situated some centimeters from the extraction system (downstream) is responsible for the incorporation of metals (mainly Fe). As it was mentioned in Sect. 3.1, the use of a high U_{acc} leads to a higher divergence of the beam. Therefore, it is expected to lead also to higher sputtering of the steel lining and in turn a higher amount of Fe reaching the substrate. In Fig. 11a the effect of U_{acc} in the angular distribution of the ions leaving an aperture of the grid system are shown (simulation results). The simulation was done using the IGUN code [66]. It is observed that the increase of $|U_{acc}|$ from 0.2 to 1 kV increases the divergence of the beamlet (and in consequence of the whole beam) and leads to the nanopattern formation on Si (see AFM images on the right in Fig 11). As expected and according to the Rutherford backscattered spectrometry (RBS) measurements, the concentration of Fe is more than three times higher for $U_{\mathrm{acc}} = -1$ kV than for $U_{\mathrm{acc}} = -0.2$ kV (1.92×10^{15} at cm^{-2} and 0.55×10^{15} at cm^{-2}, respectively). The secondary ion mass spectrometry (SIMS) depth profile in Fig. 11b shows also that for $U_{\mathrm{acc}} = -1$ kV the concentration of Fe is significantly higher than for $U_{\mathrm{acc}} = -0.2$ kV and that the Fe is mostly in the first 3 or 4 nm and then the concentration decreases. According to these results, the incorporation of Fe seems to enhance the nanostructure formation under these experimental conditions.

The results from Macko et al. [67] are consistent with these observations. They found that due to the irradiation with Kr^+, $E_{\mathrm{ion}} = 2$ keV, at $\alpha \leq 45°$ no pattern evolves on pure Si surfaces, while with the co-sputtering of a stainless steel target, ripples and dots are formed.

The continuous arrival of Fe to the substrate is counteracted by its simultaneous sputtering. Shortly after the beginning of the irradiation a steady state between

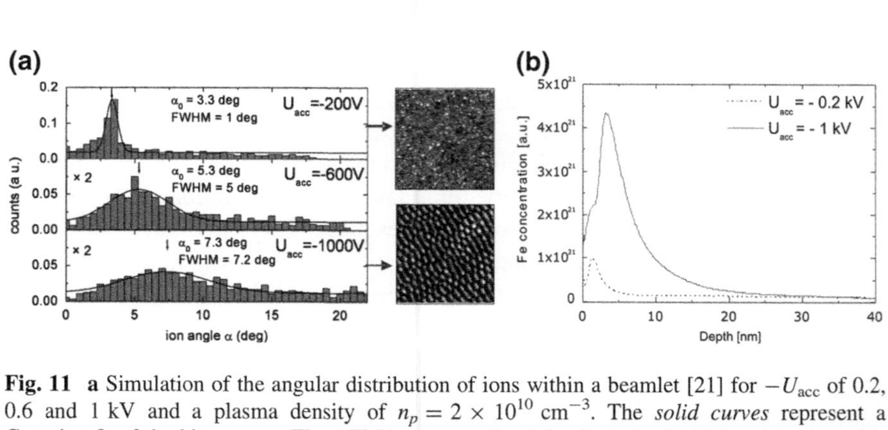

Fig. 11 **a** Simulation of the angular distribution of ions within a beamlet [21] for $-U_{\mathrm{acc}}$ of 0.2, 0.6 and 1 kV and a plasma density of $n_p = 2 \times 10^{10}$ cm^{-3}. The *solid curves* represent a Gaussian fit of the histograms. The AFM images on the right show the Si(001) topography after irradiation with Kr^+, $E_{\mathrm{ion}} = 2$ keV, $\alpha = 15°$, $\Phi = 3.4 \times 10^{18}$ cm^{-2} and U_{acc} of -0.2 kV (*top*) and -1 kV (*bottom*). The AFM images are 1×1 μm and the vertical scale 2 nm (*top*) and 6 nm (*bottom*). **b** Corresponding Fe concentration depth profile (SIMS measurements)

these two processes is reached and the Fe concentration remains constant. Thus, the sputtering of Fe has to be also considered. It is known that the sputter yield is highly affected by the incidence angle. In Fig. 12a the Fe concentration on Si(001) after the bombardment with 1 keV Kr^+, $U_{acc} = -1$ kV and -0.2 kV, $\Phi = 7.8 \times 10^{17}$ cm^{-2}, and $\alpha = 0$–$75°$ is presented. As shown above, for the experimental setup used for this study, U_{acc} represents a way to control the amount of Fe reaching the substrate. This fact is observed in the plot in Fig. 12a; the Fe concentration is significantly higher when U_{acc} is varied from -0.2 to -1 kV. Besides, the decrease of the Fe concentration with the beam incidence angle is observed, increasing slightly for $U_{acc} = -1$ kV at grazing angles. To understand this tendency the incidence angle dependence of the sputter yield (Y) should be considered. In Fig. 12b this dependence is shown, using the sputter yield values calculated with TRIM.SP code [68]. Both Y_{Fe} and Y_{Si} increase when the incidence angle increases, reaching a maximum and then decreasing again at grazing angles. Y_{Fe} reaches the maximum near $\alpha = 65°$ and Y_{Si} near $\alpha = 70°$. The dashed line in the plot shown in Fig. 12b indicates the ratio of the Y_{Fe} to Y_{Si}. The plot shows that Y_{Fe} is higher than Y_{Si} from 0° up to ~70° and that the ratio Y_{Fe}/Y_{Si} decreases continuously in this incidence angle range. At $\alpha > 70°$ Y_{Fe} is lower than Y_{Si}. These observations would explain the tendency shown in the plot in Fig. 12a. The variation of the Fe concentration in the steady state is also consistent with the model for surfactant sputtering from H. Hofsäss and K. Zhang [69]. Hofsäss and Zhang studied the ion sputtering with simultaneous co-sputtering, which they called *surfactant sputtering*. They combined different substrates and surfactants and proposed a simple model to explain steady state coverage of a substrate with surfactant atoms. The experimental observations and the model can be found in [69, 70].

The way Fe enhances the pattern formation is still unclear. The co-deposited metal atoms may affect the energy distribution of the ion collision and the sputter yield locally. It was suggested that the sputtering introduces stress to the surface

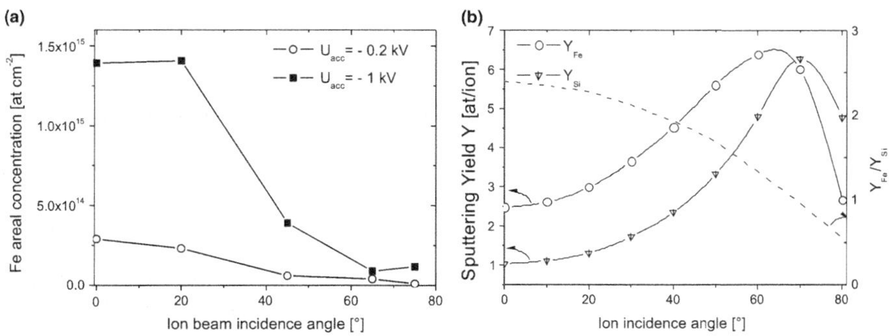

Fig. 12 **a** Ion beam incidence angle vs. Fe areal concentration (determined by RBS) for Si(001) after bombardment with 1 keV Kr^+, $\Phi = 7.8 \times 10^{17}$ cm^{-2}, $\alpha = 0$–$75°$ and $U_{acc} = -1$ kV and -0.2 kV. **b** Incidence angle dependence of Fe and Si sputter yields, calculated with TRIM.SP code [68]. The discontinuous line indicates the ratio Y_{Fe}/Y_{Si} variation with ion incidence angle

and its relieve leads to the pattern formation [71]. However, we found that no surface strain was introduced by sputtering [72].

A HRTEM cross-sectional view of the surface together with AFM images of Si(001) after the irradiation with Kr^+ at $\alpha = 20$ and 65°, is presented in Fig. 13. In both cases ripples perpendicular to the ion beam projection evolved, having higher regularity those formed at $\alpha = 20°$ (Fig. 13c, d). The wavelength and amplitude of the ripples at $\alpha = 65°$ are ~55–60 nm and ~2–3 nm, respectively, while the ripples at $\alpha = 20°$ have a wavelength ~45 nm and an amplitude ~2 nm.

As it was shown above, the Fe concentration in the steady state is highly affected by the ion beam incidence angle. According to previous results, the Fe concentration is about seven times higher for $\alpha = 20°$. In the HRTEM micrographs the crystalline Si substrate is observed for both samples and on top of it a layer is distinguished. This layer corresponds to the amorphous Si, which is formed due to the ion sputtering, together with SiO_2 formed after the exposure to

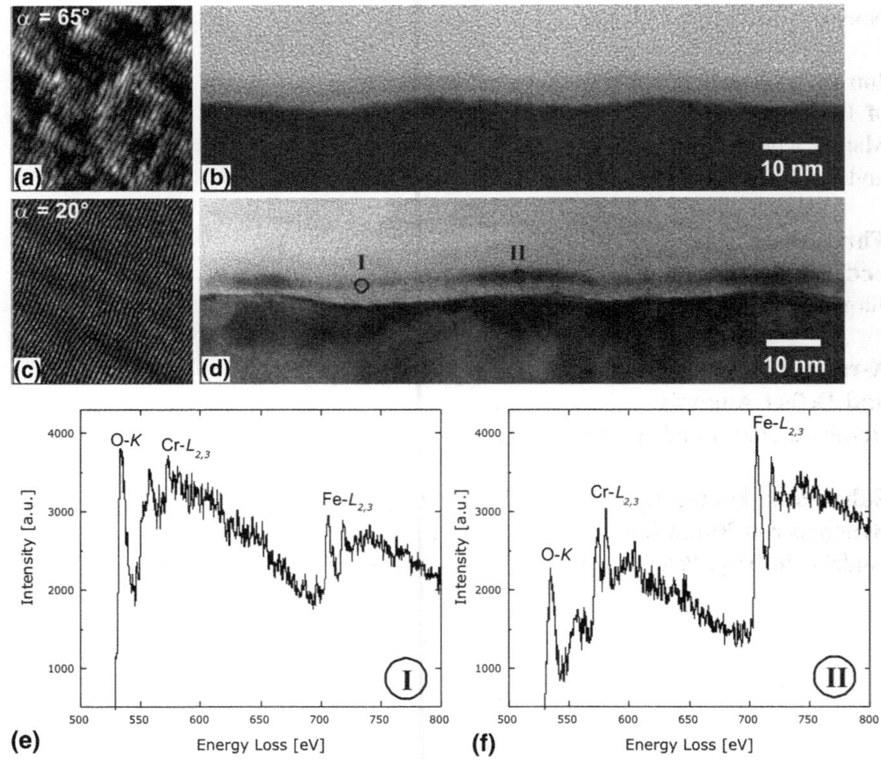

Fig. 13 a–d AFM and HRTEM images of Kr^+-sputtered Si(001) ($\Phi = 8.7 \times 10^{18}$ cm^{-2}, $U_{acc} = -1$ kV, $E_{ion} =$ **a**, **b** 2 keV, **c**, **d** 1.5 keV, $\alpha =$ **a**, **b** 65°, **c**, **d** 20°). The AFM images are 2×2 μm^2 and the Z scale is **a** 6 nm, **c** 3 nm. **e**, **f** EELS spectra corresponding to the positions I and II indicated in **d**

air. Also most contaminants are found in this layer. It is clearly observed that the layers of both samples look different. For the sample sputtered at $\alpha = 65°$ (Fig. 13b) the layer looks homogenous while the layer of the sample corresponding to $\alpha = 20°$ (Fig. 13d) has some dark regions in the crests of the ripples. The composition of the dark regions at the crests of the ripples was analyzed using electron energy loss spectroscopy (EELS) and compared with the composition at the valleys. In Fig. 13e and 13f the EELS spectra for the positions I and II indicated in Fig. 13d are shown. It is clearly seen that the concentration of Fe and Cr is higher at the crest of the ripple (dark region) than at the valley.

According to the experimental observations presented in this sub-section, pattern formation on Si at near normal ion beam incidence seems to be inherently associated with Fe incorporation.

3.6 Other Experimental Parameters

The influence of the substrate temperature in the pattern formation has been also studied. Again, a comparison of the experimental observations reported is not simple due to the many experimental parameters involved. As example the observations from Erlebacher et al. [17] could be mentioned. In agreement with the BH model which predicts that the ripples wavelength increases with temperature, they observed that increasing the temperature from 460 to 600°C the ripples wavelength increases on Si(001) bombarded with 0.75 keV Ar^+ at $\alpha = 67.5°$. Gago et al. [51], on the other hand, who studied the influence of the temperature in the dot formation on Si(001) by sputtering with 1 keV Ar^+ at normal incidence found that the pattern was not affected by the substrate temperature up to 425 K; between 425 and 525 K the dot height and wavelength decreased with the temperature and finally above 550 K no pattern evolved.

Most of the experimental observations agree that in the case of Si, the temperature shows no influence when the substrate temperature is low. At low temperatures the Si surface is amorphized by the ion bombardment while at high temperature it remains crystalline. The transition from one case to the other depends on the sputter conditions. For the bombardment of Si(001) with 1 keV Ar^+, the transition was observed between 400 and 500°C [50].

4 Other Materials

The ion beam sputtering as a patterning technique can be applied to a wide variety of materials beside Si. Here some examples will be given.

4.1 Germanium

So far, there are only few reports about the pattern formation by low-energy ion beam sputtering on Ge surfaces [18, 28, 53]. In Fig. 14 the topography of Ge(001) after irradiation with 2 keV Xe^+ at ion beam incidence angle α of 0, 5, and 10° is shown. Dots with poor lateral ordering were formed at $\alpha = 0°$ (Fig. 14a), while at $\alpha = 5°$ perpendicular-mode ripples evolved (Fig. 14b). Increasing the ion beam incidence angle, a transition to dots was observed and at angles above 30° no pattern evolved. A mixture of perpendicular-mode ripples and dots was observed at $\alpha = 10°$ (Fig. 14c). Also Carbone et al. [18] observed the simultaneous presence of ripples and dots after the irradiation of Ge(001) with 1 keV Xe^+ at $\alpha = 10°$. As in the case of other semiconductor materials, Ge is amorphized due to the ion sputtering at room temperature. Chason et al. [28] studied the irradiation of Ge(001) with 1 keV Xe^+ and observed that above 250°C the surface remains crystalline and the topography evolution changes.

Due to the similarity in the pattern formation of Si and Ge, the potential role of Fe in the topography evolution of Ge should be investigated.

4.2 III–V Compounds

The ion induced self-organization on III/V semiconductors has been investigated. The studies have shown that in many cases there is a preferential sputtering, due to the different sputtering yields of the components, which leads to the enrichment of the surface with one component. Thus, the ion sputtering of InP and GaAs results in an enrichment of In and Ga, respectively [73, 74]. The preferential sputtering makes the understanding of the processes involved even more difficult than for elementary materials.

With respect to the nanostructures formed by ion sputtering, one interesting example is the nanodots on GaSb formed at normal incidence by low energy Ar^+ sputtering. Facsko et al. [75] produced nanodots on GaSb with hexagonal ordering

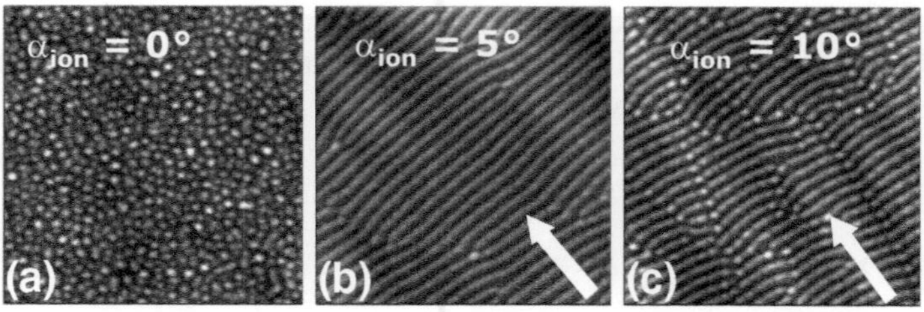

Fig. 14 AFM images Ge(001) after irradiation with Xe^+, $E_{ion} = 2$ keV, $\Phi = 6.7 \times 10^{18}$ cm^{-2}, $U_{acc} = -1$ kV, $\alpha =$ **a** 0°, **b** 5° and **c** 10°. The AFM images are 1.5 × 1.5 μm and the vertical scale **a** 8 nm, **b** 12 nm, and **c** 10 nm

by bombardment with 0.42 keV Ar^+ at normal incidence. They studied the effect of the fluence and they observed that in the fluence range of 4×10^{17} to 4×10^{18} cm^{-2}, the period of the dots increases from 18 to 50 nm, also de ordering increases and then the features stabilize. They showed also that the dots (or cones) keep the crystalline structure of the bulk GaSb and they are covered by a 2 nm amorphous layer. In a related study Facsko et al. [29] found that the wavelength of the nanodots on GaSb produced at normal incidence increases with E_{ion} (proportional to the square root of the ion energy) over a large range of energies.

Nanodots organized in hexagonally ordered domains were also observed on InP after Ar^+ bombardment but under oblique ion incidence with simultaneous sample rotation [30]. In Fig. 15 the sputtered InP surface is shown. It was irradiated at room temperature with Ar^+, $E_{ion} = 0.5$ keV, $\alpha = 40°$. Under these conditions the wavelength of the dots saturated at $\lambda \sim 85$ nm. The effect of the ion beam incidence angle under these sputter conditions was investigated. Domains of hexagonally arranged dots were formed at $\alpha \leq 50°$; with further increase of the ion beam incidence angle the periodic pattern vanished and at $\alpha = 80°$ dot formation was again observed but with smaller dot size than for $\alpha \leq 50°$. The height and wavelength of the dots at $\alpha = 40°$ increased with increasing E_{ion} (energy range from 0.35 to 1.2 keV). The sample temperature seems to be critical for InP under these conditions. In the range $T_s = 285$–375 K, the wavelength and roughness increase with temperature. The dots have the same crystalline structure as the InP bulk material, only covered by a thin amorphous layer. As mentioned above, due to preferential sputtering effects the amorphous layer shows an enrichment in In. More details about the formation of dots on InP can be found in [23, 30].

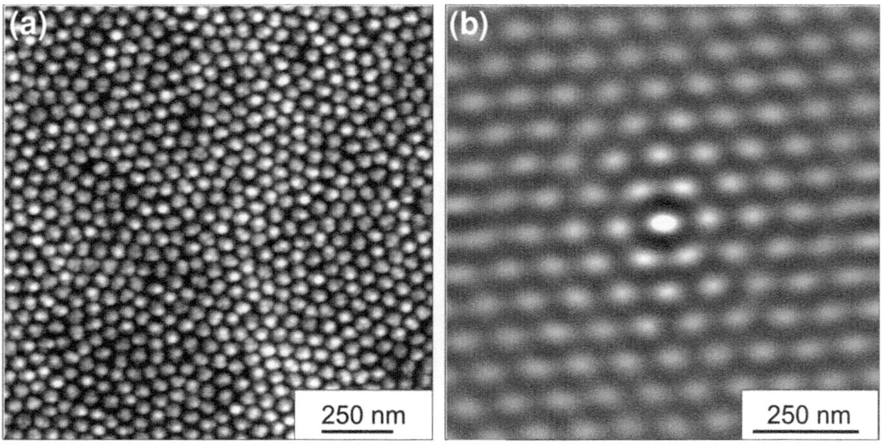

Fig. 15 a AFM image of Ar^+-sputtered InP at $\alpha = 40°$, with $E_{ion} = 0.5$ keV, $\Phi = 1.34 \times 10^{19}$ cm^{-2}. The lateral size of the image is 1.5 μm and the height scale 70 nm. **b** Two-dimensional autocorrelation function $C(r, t)$ from an enlarged region of image **a** (image size 1 μm)

Self-organized patterns can be also produced on InAs and InSb by ion beam sputtering [42].

4.3 Fused Silica

Our group has also investigated the ion beam sputtering of SiO_2. In Fig. 16 the effect of the ion beam incidence angle when SiO_2 is bombarded with Ar^+, $E_{ion} = 1.2$ keV, $U_{acc} = -1$ kV, $\Phi = 6.7 \times 10^{18}$ cm^{-2} is shown. It is observed that under these conditions the surface remains smooth at $\alpha < 40°$ (Fig. 16 I), perpendicular-mode ripples form between $\alpha \sim 40°$ and 70° (Fig. 16 II) and at $\alpha \sim 80°$ the orientation of the features is rotated by 90° (Fig. 16 III). In agreement to these results Mayer et al. [40] observed the formation of ripples oriented perpendicular to the ion beam direction after 1 keV Xe^+ bombardment at $\alpha = 55°$. Perpendicular-mode ripple formation on SiO_2 was also observed after

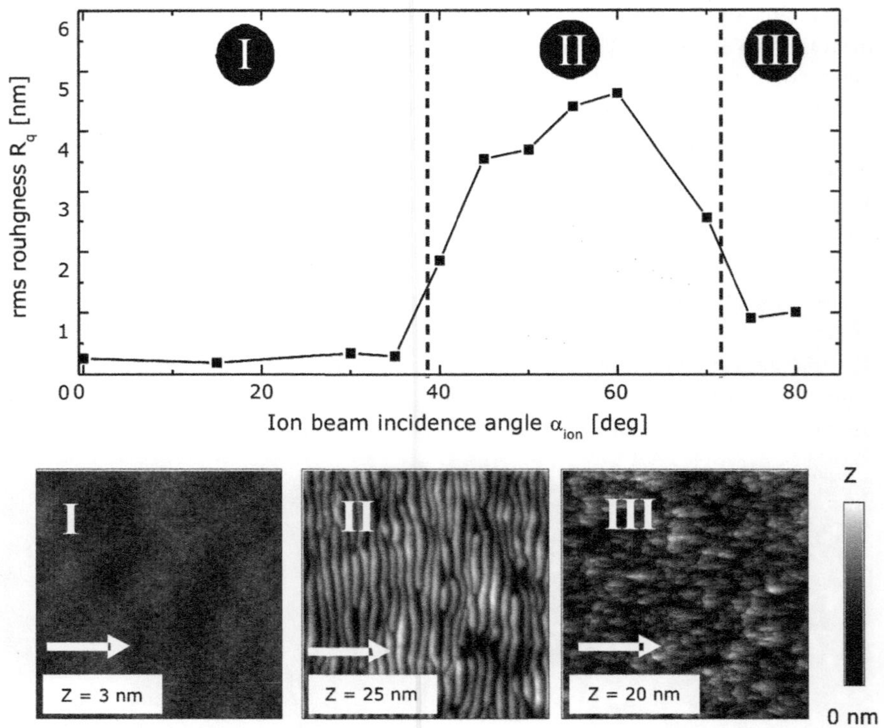

Fig. 16 Surface roughness vs. ion beam incidence angle for SiO_2 bombarded with Ar^+ ions, $E_{ion} = 1.2$ keV, $U_{acc} = -1$ kV, $\Phi = 6.7 \times 10^{18}$ cm^{-2}, without sample rotation. 2 × 2 μm AFM images of the different topographies: I: smooth surface, II: perpendicular-mode ripples, III: parallel-mode ripples. The *arrows* indicate the projection of the ion beam on the surface

bombardment with 1 keV Ar^+ at $\alpha = 45°$ [39]. Toma et al. [19] studied the pattern formation on SiO_2 due to bombardment with 0.8 keV Ar^+ and found that perpendicular- and parallel-mode ripples were formed at $\alpha = 35$ and 75°, respectively.

The temporal evolution of the ripples was studied and the increase of ripples wavelength and amplitude with the fluence was observed [19, 38, 40]. The ripples size (wavelength and amplitude) also increases with increasing E_{ion} [38–40]. On the contrary, no correlation between the wavelength and the ion flux was observed [38]. The effect of the temperature in the ripples wavelength was found to be negligible when T < 200°C while an Arrhenius-like increase of the wavelength at higher T was observed [39].

4.4 Metals

The response of crystalline metals to the ion bombardment differs from that of semiconductors and amorphous materials. The difference in the behaviour is mainly due to the higher diffusivity in metals and the non-directional character of the metallic bonds. Generally, the ion bombardment of metals does not imply amorphization. Thus, the behaviour can not be explained by Bradley and Harper model. In crystalline metals an additional destabilizing mechanism that may represent a patterning driving force has to be considered, i.e. the effect of the Ehrlich–Schwoebel (ES) energy barriers.

The ES barriers oppose the diffusion toward downhill direction. Depending on the experimental conditions the nanostructure orientation on monocrystalline metals can be given by the crystallographic orientation (when diffusive regime dominates) or the direction of the ion beam (at grazing incidence angles and low substrate temperature to enhance the erosion regime). Mounds, parallel- and perpendicular-mode ripples can be produced. On polycrystalline metals the grain boundaries do not hinder the formation of coherent ripple patterns if the erosive regime dominates. Numerous examples and a complete description of the ion-beam sputtering on metals can be found in [33, 37, 76–79].

5 Summary

This chapter focuses on the self-organized pattern formation by ion beam sputtering. The description of the technique and selected experimental observations were given. The effect of some of the several experimental parameters involved in the pattern evolution was presented, setting the main focus on the patterning of silicon surfaces. Among the numerous kind of nanostructures that can be produced, two features of great interest are ripples and dots, which under certain conditions may have a high regularity. It was shown that the angle of incidence of the ions plays a crucial role in the pattern formation, leading to different types of structures. As the surfaces become amorphous short after the sputtering begins, the effect of the crystallographic orientation is lost, at least at low temperature. The orientation of anisotropic structures is then given by the direction of the ion beam. By rotation of the substrate during sputtering isotropic patterns can be produced. Experimental results about the effect of the ion energy, substrate temperature and fluence were shown. The effect of the incorporation of metals during sputtering was also discussed. This additional factor involved in the pattern formation has been ignored for long time. Here, evidence of the importance of iron in the ripple formation on silicon was presented. Finally, examples of the application of this technique to semiconductor compounds, germanium, silica and crystalline metals were given.

According to the information presented in this chapter, the self-organized patterning by ion beam sputtering may represent an alternative to the "conventional" lithographic techniques to produced patterned surfaces. This "unconventional" bottom-up approach enables the production of nanostructures in only one step on a wide variety of materials. By the use of broad-beam ion sources the technique may be suitable for industrial use. However, as there are still many open questions about the phenomenon, further investigation should be carried out to achieve a better understanding of the processes involved. This would enable a better control of the topography evolution by the selection of appropriate experimental conditions. In particular, the role of the metal incorporation should be further investigated.

Acknowledgments The authors would like to thank the contribution of Dietmar Hirsch (SIMS measurements), Prof. Dr. Thomas Höche (TEM), Dr. Dieter Flamm (assistance with the ion source), Dr. Jürgen Gerlach (RBS), and Christoph Meinecke (RBS) from IOM. We thank also Sven Macko and Prof. Dr. Thomas Michely from the University of Cologne for the stimulating discussions. The work is financially supported by the Deutsche Forschungsgemeinschaft through FOR-845.

References

1. Navez, M., Sella, C., Chaperot, D.: Microscopie electronique-étude de lattaque du verrepar bombardement ionique. CR Acad. Sci. **254**, 240 (1962)
2. Bradley, R.M., Harper, J.M.E.: Theory of ripple topography induced by ion bombardment. J. Vac. Sci. Technol. A **6**(4), 2390–2395 (1988)
3. Castro, M., Cuerno, R., Vázquez, L., Gago, R.: Self-organized ordering of nanostructures produced by ion-beam sputtering. Phys. Rev. Lett. **94**, 016102 (2005)
4. Cuerno, R., Makse, H.A., Tomassone, S., Harrington, S.T., Stanley, H.E.: Stochastic model for surface erosion via ion sputtering: dynamical evolution from ripple morphology to rough morphology. Phys. Rev. Lett. **75**, 4464 (1995)
5. Facsko, S., Bobek, T., Stahl, A., Kurz, H., Dekorsy, T.: Dissipative continuum model for self-organized pattern formation during ion-beam erosion. Phys. Rev. B **69**, 153412 (2004)
6. Makeev, M.A., Cuerno, R., Barabási, A.-L.: Morphology of ion-sputtered surfaces. Nucl. Instrum. Methods Phys. Res. B **197**, 185–227 (2002)

7. Park, S., Kahng, B., Jeong, H., Barabási, A.-L.: Dynamics of ripple formation in sputter erosion: nonlinear phenomena. Phys. Rev. Lett. **83**, 4 (1999)
8. Rost, M., Krug, J.: Anisotropic Kuramoto-Sivashinsky equation for surface growth and erosion. Phys. Rev. Lett. **75**, 3894–3899 (1995)
9. Vogel, S., Linz, S.: Continuum modeling of sputter erosion under normal incidence: Interplay between nonlocality and nonlinearity. Phys. Rev. B **72**, 035416 (2005)
10. Sigmund, P.: A mechanism of surface micro-roughening by ion bombardment. J. Mater. Sci. **8**, 1545–1553 (1973)
11. Ziberi, B.: Ion beam induced pattern formation on Si and Ge surfaces. PhD Thesis, Leipzig (2006)
12. Kaufman, H.R., Cuomo, J.J., Harper, J.M.E.: Technology and applications of broad-beam ion sources used in sputtering. Part I. Ion source technology. J. Vac. Sci. Technol. **21**, 725–736 (1982)
13. Cuomo, J.J., Rossnagel, S.M., Kaufman, H.R.: Handbook of Ion Beam Processing Technology. Noyes Publications, USA (1989)
14. Ludwig, F.J., Eddy, C.R.J., Malis, O., Headrick, R.L.: Si(100) surface morphology evolution during normal-incidence sputtering with 100–500 eV Ar^+ ions. Appl. Phys. Lett. **81**, 2770–2772 (2002)
15. Gago, R., Cuerno, R., Varela, M., Ballesteros, C., Albella, J.M.: Production of ordered silicon nanocrystals by low-energy ion sputtering. Appl. Phys. Lett. **78**, 3316–3318 (2001)
16. Gago, R., Vázquez, L., Cuerno, R., Varela, M., Ballesteros, C., Albella, J.M.: Nanopatterning of silicon surfaces by low-energy ion-beam sputtering: dependence on the angle of ion incidence. Nanotechnology **13**, 304–308 (2002)
17. Erlebacher, J., Aziz, M.J., Chason, E., Sinclair, M.B., Floro, J.A.: Spontaneous pattern formation on ion bombarded Si(001). Phys. Rev. Lett. **82**(11), 2330–2333 (1999)
18. Carbone, D., Alija, A., Plantevin, O., Gago, R., Facsko, S., Metzger, T.H.: Early stage of ripple formation on Ge(001) surfaces under near-normal ion beam sputtering. Nanotechnology **19**, 035304 (2008)
19. Toma, A., Buatier de Mongeot, F., Buzio, R., Firpo, G., Bhattacharyya, S.R., Boragno, C., Valbusa, U.: Ion beam erosion of amorphous materials: evolution of surface morphology. Nucl. Instrum. Methods Phys. Res. B **230**, 551–554 (2005)
20. Ziberi, B., Frost, F., Höche, T., Rauschenbach, B.: Ripple pattern formation on silicon surfaces by low-energy ion beam erosion: Experiment and theory. Phys. Rev. B **72**, 235310 (2005)
21. Ziberi, B., Frost, F., Neumann, H., Rauschenbach, B.: Ripple rotation, pattern transitions, and long range ordered dots on silicon by ion beam erosion. Appl. Phys. Lett. **92**, 063102 (2008)
22. Ziberi, B., Frost, F., Tartz, M., Neumann, H., Rauschenbach, B.: Importance of ion beam parameters on self-organized pattern formation on semiconductor surfaces by ion beam erosion. Thin Solid Films **459**, 106–110 (2004)
23. Frost, F., Rauschenbach, B.: Nanostructuring of solid surfaces by ion-beam erosion. Appl. Phys. A **77**, 1–9 (2003)
24. Brown, A.D., Erlebacher, J.: Temperature and fluence effects on the evolution of regular surface morphologies on ion-sputtered Si(111). Phys. Rev. B **72**, 075350 (2005)
25. Chini, T.K., Sanyal, M.K., Bhattacharyya, S.R.: Energy-dependent wavelength of the ion-induced nanoscale ripple. Phys. Rev. B **66**, 153404 (2002)
26. Carter, G., Nobes, M.J., Paton, F., Williams, J.S., Whitton, J.L.: Ion bombardment induced ripple topography on amorphous solids. Radiat. Eff. Defects Solids **33**, 65–73 (1977)
27. Zhang, K., Rotter, F., Uhrmacher, M., Ronning, C., Hofsäss, H., Krauser, J.: Pattern formation of Si surfaces by low-energy sputter erosion. Surf. Coat Technol. **201**, 8299–8302 (2007)
28. Chason, E., Mayer, T.M., Kellermann, B.K., McIlroy DN, J.H.A.: Roughening instability and evolution of the Ge(001) surface during ion sputtering. Phys. Rev. Lett. **72**(19), 3040–3043 (1994)

29. Facsko, S., Kurz, H., Dekorsy, T.: Energy dependence of quantum dot formation by ion sputtering. Phys. Rev. B **63**, 165329 (2001)
30. Frost, F., Schindler, A., Bigl, F.: Roughness evolution of ion sputtered rotating InP surfaces: pattern formation and scaling laws. Phys. Rev. Lett. **85**(19), 4116–4119 (2000)
31. Boragno, C., Buatier de Mongeot, F., Constantini, G., Molle, A., de Sanctis, D., Valbusa, U.: Time evolution of the local slope during Cu(110) ion sputtering. Phys. Rev. B **68**, 094102 (2003)
32. Valbusa, U., Boragno, C., Buatier de Mongeot, F.: Nanostructuring by ion beam. Mater. Sci. Eng. C **23**, 201–209 (2003)
33. Rusponi, S., Constantini, G., Boragno, C., Valbusa, U.: Ripple wave vector rotation in anisotropic crystal sputtering. Phys. Rev. Lett. **81**, 2735–2738 (1998)
34. van Dijken, S., De Bruin, D., Poelsema, B.: Kinetic physical etching for versatile novel design of well ordered self-affine nanogrooves. Phys. Rev. Lett. **86**, 4608–4611 (2001)
35. Toma, A., Šetina Batič, B., Chiappe, D., Boragno, C., Valbusa, U., Godec, M., Jenko, M., Buatier de Mongeot, F.: Patterning polycrystalline thin films by defocused ion beam: the influence of initial morphology on the evolution of self-organized nanostructures. J. Appl. Phys. **104**, 104313 (2008)
36. Ghose, D.: Ion beam sputtering induced nanostructuring of polycrystalline metal films. J. Phys: Condens. Matter **21**, 224001 (2009)
37. Karmakar, P., Ghose, D.: Ion beam sputtering induced ripple formation in thin metal films. Surf. Sci. **554**, L101–L106 (2004)
38. Flamm, D., Frost, F., Hirsch, D.: Evolution of surface topography of fused silica by ion beam sputtering. Appl. Surf. Sci. **179**, 96–102 (2001)
39. Umbach, C.C., Headrick, R.L., Chang, K.-C.: Spontaneous nanoscale corrugation of ion-eroded SiO_2: the role of ion-irradiation-enhanced viscous flow. Phys. Rev. Lett. **87**, 246104 (2001)
40. Mayer, T.M., Chason, E., Howard, A.J.: Roughening instability and ion-induced viscous relaxation of SiO_2 surfaces. J. Appl. Phys. **76**, 1633–1643 (1994)
41. Frost, F., Ziberi, B., Schindler, A., Rauschenbach, B.: Surface engineering with ion beams: from self-organized nanostructures to ultra-smooth surfaces. Appl. Phys. A **91**, 9 (2008)
42. Frost, F., Fechner, R., Flamm, D., Ziberi, B., Frank, W., Schindler, A.: Ion beam assisted smoothing of optical surfaces. Appl. Phys. A **78**, 651–654 (2004)
43. Frost, F., Fechner, R., Ziberi, B., Völlner, J., Flamm, D., Schindler, A.: Large area smoothing of surfaces by ion bombardment: fundamentals and applications. J. Phys.: Condens. Matter **21**, 224026 (2009)
44. Keller, A., Roßbach, S., Facsko, S., Möller, W.: Simultaneous formation of two ripple modes on ion sputtered silicon. Nanotechnology **19**, 135303 (2008)
45. Chini, T.K., Datta, D.P., Bhattacharyya, S.R.: Ripple formation on silicon by medium energy ion bombardment. J. Phys.: Condens. Matter **21**, 224004 (2009)
46. Ozaydin-Ince, G., Ludwig, K.F.J.: In situ X-ray studies of native and Mo-seeded surface nanostructuring during ion bombardment of Si(100). J. Phys.: Condens. Matter **21**, 224008 (2009)
47. Madi, C.S., Bola George, H., Aziz, M.J.: Linear stability and instability pattern in ion-sputtered silicon. J. Phys.: Condens. Matter **21**, 224010 (2009)
48. Ziberi, B., Frost, F., Höche, T., Rauschenbach, B.: Ion-induced self-organized dot and ripple patterns on Si surfaces. Vacuum **81**, 155–159 (2006)
49. Madi, C.S., Davidovitch, B., Bola George, H., Norris, S.A., Brenner, M.P., Aziz, M.J.: Multiple bifurcation types and the linear dynamics of ion sputtered surfaces. Phys. Rev. Lett. **101**, 246102 (2008)
50. Ozaydin, G., Ludwig, F.J., Zhou, H., Zhou, L., Headrick, R.L.: Transition behavior of surface morphology evolution of Si(100) during low-energy normal-incidence Ar^+ ion bombardment. J. Appl. Phys. **103**, 033512 (2008)
51. Gago, R., Vázquez, L., Plantevin, O., Sánchez-García, J.A., Varela, M., Ballesteros, C., Albella, J.M., Metzger, T.H.: Temperature influence on the production of nanodot patterns by ion beam sputtering of Si(001). Phys. Rev. B **73**, 145414 (2006)

52. Habenicht, S., Lieb, K.P., Koch, J., Wieck, A.D.: Ripple propagation and velocity dispersion on ion-beam eroded silicon surfaces. Phys. Rev. B **65**, 115327 (2002)
53. Ziberi, B., Frost, F., Rauschenbach, B.: Pattern transition on Ge surfaces during low-energy ion beam erosion. Appl. Phys. Lett. **88**, 173115 (2006)
54. Erlebacher, J., Aziz, M.J., Chason, E., Sinclair, M.B., Floro, J.A.: Nonlinear amplitude evolution during spontaneous patterning of ion-bombarded Si(001). J. Vac. Sci. Technol. A **18**, 115–120 (2000)
55. Pelliccione, M., Lu, T.-M.: Evolution of thin-film morphology: modeling and simulations. Springer, Berlin-Heidelberg (2008)
56. Hazra, S., Chini, T.K., Sanyal, M.K., Grenzer, J., Pietsch, U.: Ripple structure of crystalline layers in ion-beam-induced Si wafers. Phys. Rev. B **70**, 121307 (2004)
57. Güntherschulze, A., Tollmien, W.: Neue Untersuchungen über die Kathodenzerstäubung der Glimmentladung. Z Phys. **119**, 685–695 (1942)
58. Ma, X.L., Shang, N.G., Li, Q., Lee, I., Bello, I., Lee, S.T.: Microstructural characterization of Si cones fabricated by Ar^{+}- sputtering Si/Mo targets. J. Cryst. Growth **234**, 654–659 (2002)
59. Wehner, G.K.: Cone formation as a result of whisker growth on ion bombardment metal surfaces. J. Vac. Sci. Technol. A **3**, 1821–1835 (1985)
60. Wehner, G.K., Hajicek, D.J.: Cone formation on metal targets during sputtering. J. Appl. Phys. **42**(3), 1145–1149 (1971)
61. Morishita, S., Fujimoto, Y., Okuyama, F.: Morphological and structural features of copper seed cones. J. Vac. Sci. Technol. A **6**, 217–222 (1988)
62. Ozaydin, G., Özcan, A.S., Wang, Y., Ludwig, F.J., Zhou, H., Headrick, R.L.: Real-time X-ray of the growth of Mo-seeded Si nanodots by low-energy ion bombardment. Nucl. Instrum. Methods Phys. Res. B **264**, 47–54 (2007)
63. Ozaydin, G., Özcan, A.S., Wang, Y., Ludwig, K.F.J., Zhou, H., Headrick, R.L., Siddons, D.P.: Real-time x-ray studies of Mo-seeded Si nanodot formation during ion bombardment. Appl. Phys. Lett. **87**, 163104 (2005)
64. Sánchez-García, J.A., Gago, R., Caillard, R., Redondo-Cubero, A., Martin-Gago, J.A., Palomares, F.J., Fernández, M., Vázquez, L.: Production of nanohole/nanodot patterns on Si(001) by ion beam sputtering with simultaneous metal incorporation. J. Phys.: Condens. Matter **21**, 224009 (2009)
65. Sánchez-García, J.A., Vázquez, L., Gago, R., Redondo-Cubero, A., Albella, J.M., Czigány, Z.: Tuning the surface morphology in self-organized ion beam nanopatterning of Si(001) via metal incorporation: from holes to dots. Nanotechnology **19**, 355306 (2008)
66. Becker, R., Herrmannsfeldt, W.B.: IGUN: a program for the simulation of positive ion extraction including magnetic fields. Rev. Sci. Instrum. **63**, 3 (1992)
67. Macko, S., Frost, F., Ziberi, B., Förster, D., Michely, T.: Is keV ion-induced pattern formation on Si(001) caused by metal impurities? Nanotechnology **21**, 085301 (2010)
68. Eckstein, W.: Backscattering and sputtering with the Monte-Carlo program TRIM.SP. Radiat. Eff. Defects Solids **130**, 239–250 (1994)
69. Hofsäss, H., Zhang, K.: Surfactant sputtering. Appl. Phys. A **92**, 517–524 (2008)
70. Hofsäss, H., Zhang, K.: Fundamentals of surfactant sputtering. Nucl. Instrum. Methods Phys. Res. B **267**, 2731–2734 (2009)
71. Ozaydin, G., Ludwig, F.J., Zhou, W., Headrick, R.L.: Effects of Mo seeding on the formation of Si nanodots during low-energy ion bombardment. J. Vac. Sci. Technol. B **26**, 551 (2008)
72. Carbone, D., Biermanns, A., Ziberi, B., Frost, F., Plantevin, O., Pietsch, U., Metzger, T.H.: Ion-induced nanopatterns on semiconductor surfaces investigated by grazing incidence x-ray scattering techniques. J. Phys.: Condens. Matter **21**, 224007 (2009)
73. Nozu, M., Tanemura, M., Okuyama, F.: Direct evidence for In-crystallite growth on sputter-induced InP cones. Surf. Sci. Lett. **304**, L468–L474 (1994)
74. Tanemura, M., Aoyama, S., Fujimoto, Y., Okuyama, F.: Structural and compositional analyses of coned formed on ion-sputtered GaAs surfaces. Nucl. Instrum. Methods Phys. Res. B **61**, 451–456 (1991)

75. Facsko, S., Dekorsy, T., Koerdt, C., Trappe, C., Kurz, H., Vogt, A., Hartnagel, H.L.: Formation of ordered nanoscale semiconductor dots by ion sputtering. Science **285**, 1551–1553 (1999)
76. Buatier de Mongeot, F., Valbusa, U.: Applications of metal surfaces nanostructured by ion beam sputtering. J. Phys.: Condens. Matter **21**, 224022 (2009)
77. Toma, A., Chiappe, D., Šetina Batič, B., Godec, M., Jenko, M., Buatier de Mongeot, F.: Erosive versus shadowing instabilities in the self-organized ion patterning of polycrystalline metal films. Phys. Rev. B **78**, 4 (2008)
78. Stepanova, M., Dew, S.K.: Ion beam sputtering nanopatterning of thin metal films: the synergism of kinetic self-organization and coarsening. J. Phys.: Condens. Matter **21**, 224014 (2009)
79. Zhang, K., Uhrmacher, M., Hofsäss, H., Krauser, J.: Magnetic texturing of ferromagnetic thin films by sputtering induced ripple formation. J. Appl. Phys. **103**, 083507 (2008)

Three-Dimensional Open Cell Structures: Evaluation and Fabrication by Additive Manufacturing

Jürgen Stampfl, Heinz E. Pettermann and Mathias H. Luxner

Abstract Biological materials (e.g. wood, trabecular bone, marine skeletons) rely heavily on the use of cellular architecture, which provides several advantages: (1) The resulting structures can bear the endurable mechanical loads using a minimum of a given bulk material, thus enabling the use of lightweight design principles. (2) The inside of the structures is accessible to body fluids which deliver the required nutrients. (3) Furthermore cellular architectures can grow organically by adding or removing individual struts or by changing the shape of the constituting elements. All these facts make the use of cellular architectures a reasonable choice for nature. Using Additive Manufacturing Technologies (AMT) it is now possible to fabricate such structures for applications in engineering and biomedicine. In this book chapter we present methods which allow the 3D-analysis of the mechanical properties of cellular structures with open porosity. Various different cellular architectures are studied. In order to quantify the influence of architecture, the apparent density is always kept constant. Various lithography based AMT are described and compared regarding their suitability for the fabrication of cellular structures.

J. Stampfl (✉)
Institute of Materials Science and Technology, Vienna University of Technology, Favoritenstr. 9, 1040 Wien, Austria
e-mail: jstampfl@pop.tuwien.ac.at

H. E. Pettermann
Institute of Lightweight Design and Structural Biomechanics, Vienna University of Technology, Gusshausstrasse 27–29, 1040 Wien, Austria
e-mail: pettermann@ilsb.tuwien.ac.at

M. H. Luxner
LUXNER Engineering ZT GmbH, Rathausstr. 1, 6460 Imst, Austria
e-mail: m.luxner@luxner-zt.com

F. A. Lasagni and A. F. Lasagni (eds.), *Fabrication and Characterization in the Micro-Nano Range,* Advanced Structured Materials, 10,
DOI: 10.1007/978-3-642-17782-8_5, © Springer-Verlag Berlin Heidelberg 2011

1 Introduction

Cellular structures are one of the basic design elements of many biological materials (e.g. wood, trabecular bone, marine skeletons) [1–4]. Cellular design offers several benefits: The resulting structures can bear the endurable mechanical loads using a minimum of a given bulk material, thus enabling the use of light-weight design principles. The inside of the structures is accessible to body fluids which deliver the required nutrients. Furthermore cellular architectures can grow organically by adding or removing individual struts or by changing the shape of the constituting elements. All these facts make the use of cellular architectures a reasonable choice for nature. Using additive manufacturing technologies (AMT) it is now possible to fabricate such cellular structures with almost arbitrary complexity. Direct [5–9] as well as indirect [10, 11] AMT have been used to fabricate cellular structures, whose most appealing application is the use as scaffold for biomedical engineering. Cellular structures made by AMT are depicted in Fig. 1.

The final goal for applications in biomedicine is the fabrication of structures with defined structural as well as functional properties. Biologically inspired materials offer a large potential to tailor these properties. Regarding structural

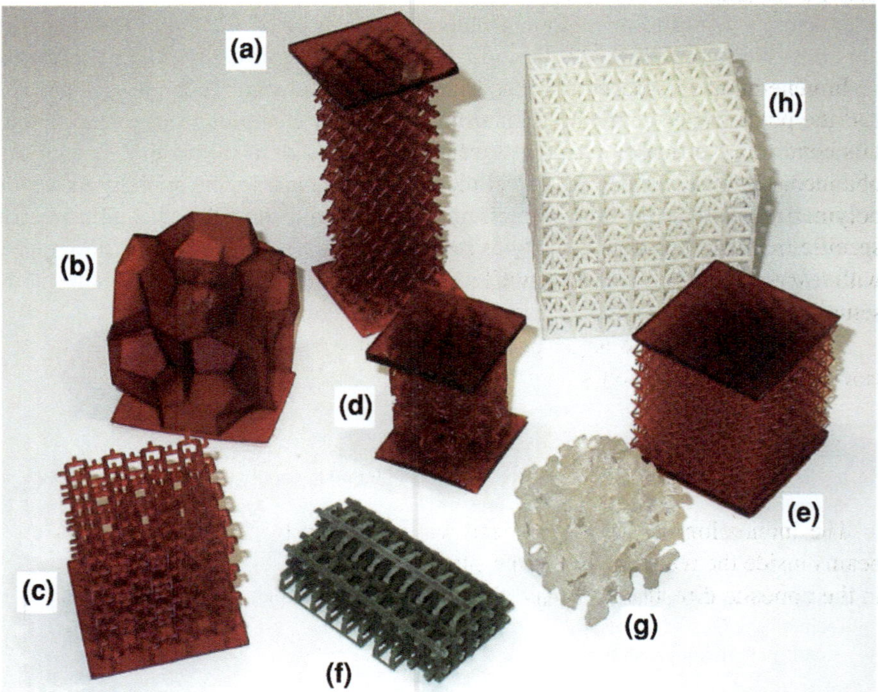

Fig. 1 Sample structures fabricated by various Rapid Prototyping processes. **a–e** Fabricated by Digital Light Processing, **f** manufactured by wax printing, **g–h** fabricated by Selective Laser Sintering

properties, biological cellular structures exhibit a significantly more irregular architecture compared to cellular engineering structures (e.g. truss structures). The role of irregularity in biological materials is not evident. Irregularity in natural biomaterials cannot be justified by a lack of "manufacturing skills" since many biomaterials (e.g. the skeleton of radiolaria, Fig. 2b, or the skeleton of deep-sea sponge euplectella [1]) exhibit a very regular and well defined micro-architecture. Nevertheless, the micro-architecture of most cellular biomaterials (e.g. cancellous bone, see Fig. 2a, which shows a cross-section through a human femur) is not completely regular.

On the level of micro-architecture cancellous bone is made up of an interconnected network of rods or plates (*trabeculae*). The trabecula are mainly oriented along the trajectories of applied mechanical stress. The trabecula along the stress direction are connected by another class of trabecula which is oriented perpendicular to the stress-direction in order to mechanically stabilize the cellular network. As can be seen from Fig. 2a there is a significant degree of irregularity on the level of micro-architecture.

In [12, 13] studies on the overall nonlinear response, of regular and disordered Simple Cubic and Kelvin structures are presented.

The influence of different kinds of structural imperfections on the effective mechanical properties of two- and three-dimensional cellular solids has been widely investigated. [14] investigate the effects of non-periodic microstructure and missing cell walls on elastic moduli, plastic collapse strength, and localization of deformation of two-dimensional cellular solids by means of finite element modelling (FEM). [15] employ the FEM to study the localization behaviour of two-dimensional honeycombs with and without defects.

Regular and disordered, low-density two-dimensional open-cell foam models subjected to large deformations are investigated numerically by [16]. In [17] the authors utilize the Voronoi tessellation technique and the FEM to analyse the effect of disorder in cell shapes and non-uniform cell wall thickness on the elastic properties of two-dimensional honeycombs.

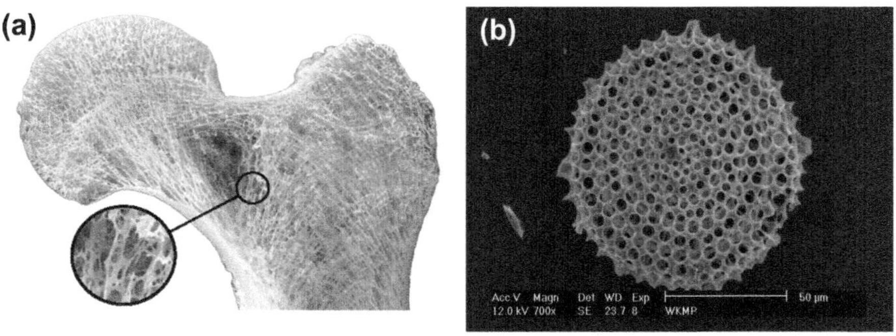

Fig. 2 Biomaterials like trabecular bone (**a**) or diatoms (**b**) are based on cellular architecture

In the case of three-dimensional structures mostly Voronoi foams are utilized. Ref. [18] investigate the elastic behaviour of three-dimensional random foams using the FEM. The effect of cell disorder on the elastic properties of three-dimensional, low-density, open-cell Voronoi foam models is analysed by Zhu [19], where the high strain compression of low-density three-dimensional open-cell polymer FEM foam models are investigated. The mechanical behaviour of linear elastic open cell foams is investigated by [20] using three-dimensional FEM Voronoi models.

2 Finite Element Modelling

2.1 Methods

All numerical investigations are carried out by means of the Finite Element package ABAQUS/Standard (Version 6.5.3, ABAQUS, Inc., Providence, RI). Beam element based models with a relative density of 12.5% are utilized throughout this section.

2.1.1 Base Cells

For this work six generic three-dimensional structures are selected, attempting to choose various topologies with a variety of mechanisms governing their behaviour. At first, all of them show regular geometries which are repeated periodically in all principal directions. The smallest periodic unit of a structure will be called "base cell" in the present work. The six different base cells to be investigated are shown in Fig. 3. They comprise Simple Cubic (SC), Gibson Ashby (GA), Reinforced Body Centred Cubic (RBCC), and Body Centred Cubic (BCC) as introduced in [21] as well as Kelvin (KV) and Weaire Phelan (WP) structures. Each structure exhibits a relative density of 12.5% and consists of struts with circular cross-sections with constant diameter. The dimensions of all base cells are $4 \times 4 \times 4$ mm. All base cells possess cubic material symmetry.

2.1.2 Finite Sample Models

Out of the six generic structures the Simple Cubic (SC) structure is selected on account of its distinct anisotropy and its various deformation modes in different directions. Since localization of deformation must not be excluded as a deformation mode, unit cell models are no longer sufficient. To overcome the limits of unit cell models [12], i.e. to allow for arbitrary localization patterns, finite structures are modelled.

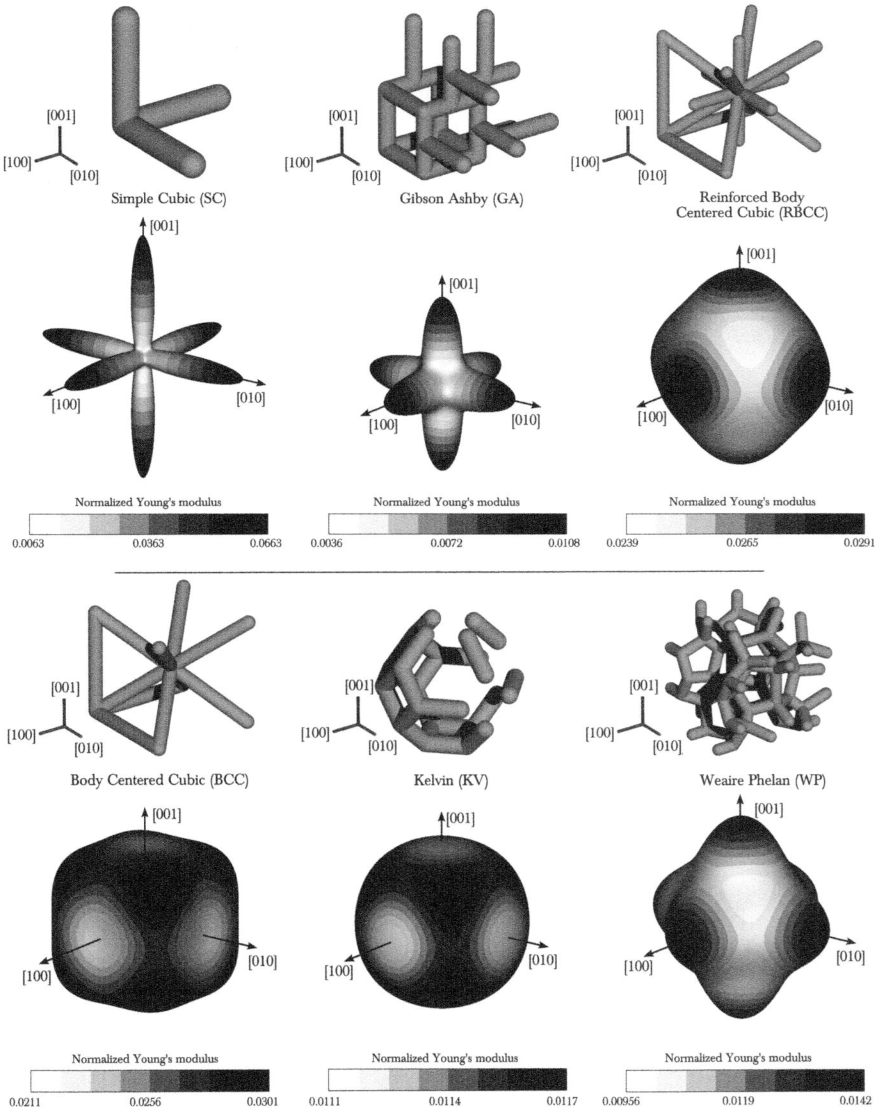

Fig. 3 Base cells of the investigated structures with a relative density of 12.5% (in the *upper* of each row) and their individually scaled normalized Young's moduli in all spatial directions (*below*). Note the different scalings of the representations of the normalized Young's moduli

Cuboidal samples with lattice orientations [001], [021], [011], and [111] are generated by arranging accordingly rotated base cells. The dimensions of the cuboidal samples are approximately $32 \times 32 \times 32\,\mathrm{mm}$.

2.2 Linear Elastic Behaviour

The linear elastic properties of the structures presented above are investigated by a periodic microfield approach. Based on three-dimensional periodic unit cell models the entire elastic tensors are predicted for six different generic structures with regular geometries. The anisotropic stiffness and the directional sensitivities are presented. Out of these six structures the two with the highest and the lowest degree of anisotropy are selected for further studies. On large periodic unit cells geometrical disorder of varying level is introduced to the structures and their effect on the linear elastic behaviour is studied. Second, the nonlinear mechanical behaviour is investigated in detail for the two selected structures. Since this, in general, cannot be accomplished by means of periodic unit cell models, large finite samples are employed which are loaded by uniaxial compression. Different orientations of the structural lattice with respect to the loading direction are realized. As for the linear cases, regular structures are considered first and then a systematic study on the influences of structural disorder is performed. On the one hand, the focus is set to the overall nonlinear response, in particular the peak load and the overall stress–strain behaviour in regimes moderately beyond the load maximum. On the other hand, the influence of structural disorder on the spatial deformation distribution is studied, i.e. whether or not deformation localization occurs and to which degree.

The anisotropy of the linear elastic stiffness of the regular structures is investigated for all six cases. Periodic unit cell models of single base cells are employed and the required number of independent load cases is solved for each structure. From these responses the entire elastic tensors are assembled. The Young's moduli in all spatial directions are extracted by rotational transformations of the tensors. The three-dimensional visualizations of the normalized Young's moduli E^*/E_s of the structures are given in Fig. 3. E_s refers to the Young's modulus of the bulk material. Note that for better visualization the scale of the normalized Young's moduli's contours is different for different structures. The normalized Young's moduli for the orientations [001], [021], [011], and [111] are listed in Table 1. Since all structures show cubic material symmetry, the extremal stiffness

Table 1 Predicted normalized Young's moduli, E^*/E_s, of the regular Simple Cubic (SC), Gibson Ashby (GA), Reinforced Body Centered Cubic (RBCC), Body Centered Cubic (BCC), Kelvin (KV), and Weaire Phelan (WP) structures for a relative density of 12.5% at selected directions

	$E^*/E_s \times 10^{-2}$			
	[001]	[021]	[011]	[111]
SC	6.630	1.193	0.816	0.631
GA	1.080	0.551	0.432	0.360
RBCC	2.906	2.632	2.499	2.389
BCC	2.106	2.460	2.716	3.007
KV	1.109	1.140	1.158	1.174
WP	1.421	1.152	1.041	0.956

values appear in the [001] and the[111] direction [22]. The SC structure shows the most pronounced anisotropy and directional sensitivity with respect to the Young's modulus. It exhibits very stiff behaviour in the principal directions and a strong decrease of the stiffness apart the principal directions. In contrast, the KV structure is the most isotropic one and exhibits nearly equal values for the normalized Young's modulus in all directions. The maximum occurs for the [111] direction. Because of the significant difference in their anisotropy the SC structure and the KV structure are chosen for subsequent investigations. From this analysis it can be concluded that the elastic response of the structure to loads in different directions can be tailored to a large extent by choosing specific base cell designs.

2.2.1 Disordered Structures

Within this work the term disorder is used in terms of randomly shifted vertex positions, and that in terms of the topology the structures are still ordered. In [13] the influence of other forms of irregularity (e.g. missing struts) is investigated in detail.

For the following investigation the vertices of the regular geometries are shifted to random positions by a fixed distance, δ. For the shifting direction a spatially random distribution is adopted. The shifting distance, δ, is expressed in fractions of the strut length l of a regular reference structure. Note that all struts of regular SC and KV structures exhibit equal length. Displacement magnitudes $\delta/l = 1/16, 1/8, 1/4$, and $3/8$ are realized, see Figs. 4 and 5, upper rows. Such disorder

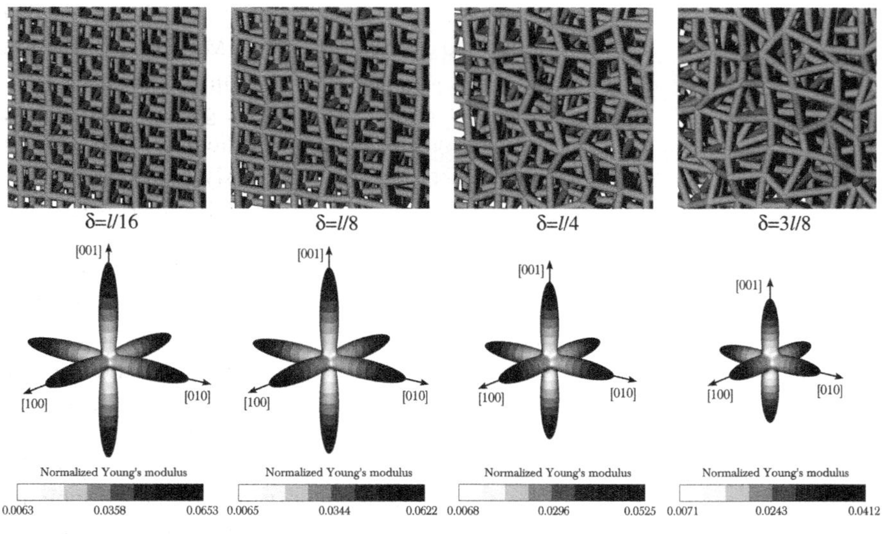

Fig. 4 Influence of structural disorder, δ, on the elastic behaviour of the Simple Cubic structure with a relative density of 12.5%; details of the disordered structures (*top*) and the predicted, equally scaled, normalized Young's moduli in all directions (*bottom*)

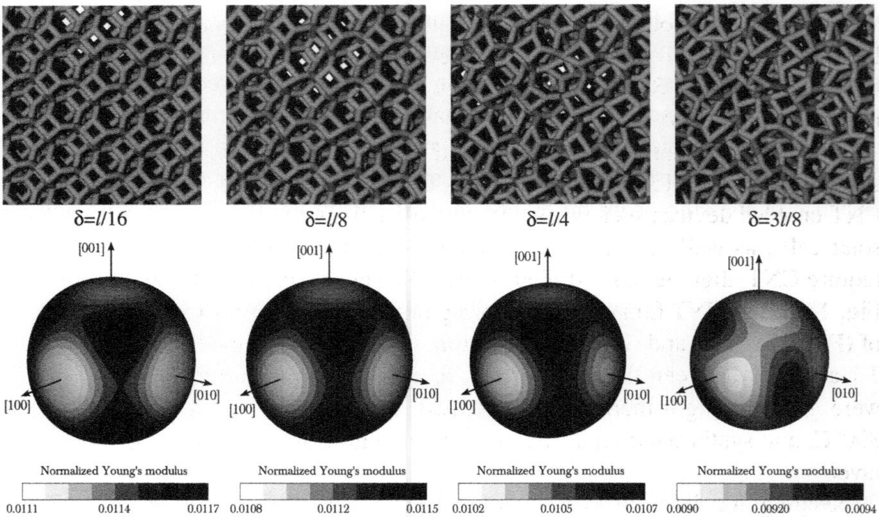

Fig. 5 Influence of structural disorder, δ, on the elastic behaviour of the Kelvin structure with a relative density of 12.5%; details from the disordered structures (*top*) and the predicted, equally scaled, normalized Young's moduli for all directions (*bottom*)

in the structures, however, increases the strut lengths and, consequently, affects the density. Thus, for representing the desired density the strut diameter is adapted accordingly. Because of the random character of the structural disorder, single base cells are no longer sufficient. Still, unit cell models are employed but consist now of some $8 \times 8 \times 8$ base cells. For easier application of periodic boundary conditions the vertices located at the unit cell boundaries remain unchanged. Four different degrees of disorder are analysed. For each of them, five different models (having the same statistical descriptors but different discrete realizations) are generated and the elastic tensors are predicted.

Figures 4 and 5 show details from the disordered structures (top) and the normalized Young's moduli for all directions (bottom). The latter shows one out of the five corresponding models for each level of disorder. The contour scaling is individual for each plot but the length scaling is the same for all plots in a figure.

In the case of the SC structure it can be seen that with increasing disorder the normalized Young's moduli in the principal directions decrease, whereas for the other directions the normalized Young's moduli increase. A pronounced anisotropy remains even for the most disordered case. The directions of the extremal values are unchanged compared to the regular structure. The KV structure shows a decrease of the normalized Young's moduli for all directions. Note that for the most disordered case the Young's moduli are nearly uniform with respect to the direction, i.e. isotropy is approached closely, and that the extremal values are no longer aligned with the principal directions of the regular structure.

2.3 Nonlinear Behaviour: Deformation Localization

Using the methods described in the previous section, the stiffness of cellular scaffolds can be estimated. In the case of scaffolds for tissue engineering it is therefore possible to adjust the stiffness of the scaffold to the stiffness of the surrounding tissue, which is beneficial for the integration of the artificial structure into the biological environment. For determining the maximum load which a cellular structure can bear, the elastic models have to be expanded to the nonlinear regime. Special attention has to be paid to the onset of deformation localization which frequently serves as the starting point of structural failure.

So far the linear elastic behaviour was considered for which periodic unit cells (of different sizes) are appropriate tools. They can also handle nonlinear deformations to a certain extent, however, other approaches have to be adopted for modelling deformation localization. The following investigations are based on finite samples which are loaded by uniaxial compression. The loading directions are given with respect to the lattice orientations and for disordered cases with respect to the lattice of the regular parent structures from which they are derived.

Again, the SC and the KV structures are studied. The results are given in terms of selected plots of deformed structures and histograms showing the total energy distribution in the models. The presented results are restricted to the deformation regime before the onset of contact between struts.

In Fig. 6 (first column) the deformation patterns of the regular SC structures at four different lattice orientations are shown without scaling the deformations. Pronounced localization of deformation can be seen. For the load cases [001] and [021] the localization takes place in a principal structure plane extending over one and two layers, respectively. These deformation patterns, however, are only one feature of the localization. The second parameter field required for fully describing the localization are the components of displacements within that plane, i.e. the direction into which the struts tilt. This will be called the localization displacement in the following. For the load cases [001] and [021] the localization displacements occur along a principal structure direction. In the [011] orientation the localization takes place in two perpendicular planes forming an "X", also showing localization displacements in principal directions. Note that, here, localization is triggered at the edge where the free and the fixed faces meet, so that other sample sizes may give rise to different deformation patterns. For the regular [111] case no distinct localization can be seen. The deformation concentrates in (011) planes but extends over the entire region which is not affected by the top and bottom constraints (representing rigid plates).

The overall stress–strain curves corresponding to the above scenarios are shown in Fig. 7 by solid lines. In all cases plastic yielding starts well before deformation localization sets in. For [001] compression the highest peak load is reached at the smallest overall strain followed by a sharp drop upon localization. In [021] and [011] loading the behaviour is similar, but with markedly reduced peak values at higher overall strains and only a moderate load reduction upon localization.

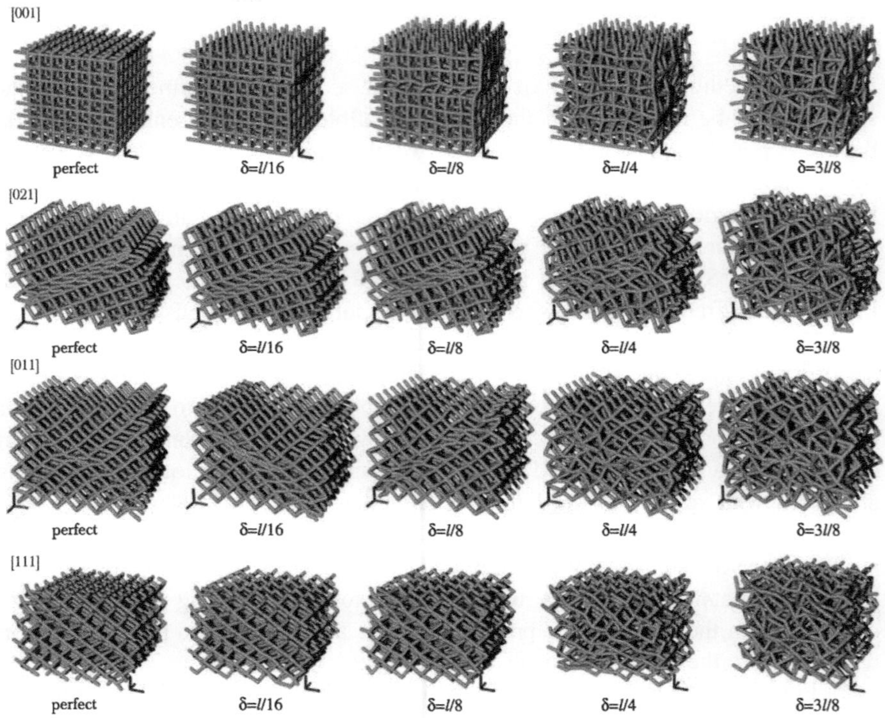

Fig. 6 Predicted deformation patterns (without deformation scaling) of Simple Cubic finite structures under uniaxial vertical compression for increasing disorder, δ, (from *left* to *right*) and for lattice orientations [001], [021], [011], and [111] (from *top* to *bottom*)

Finally, the [111] case shows the lowest peak load and almost no subsequent decrease at high overall strains.

The behaviour of the regular KV structures for the selected orientations is given in Fig. 8 by the solid lines. All predictions are very similar both with respect to the peak loads and the corresponding overall strains as well as with respect to the rather smooth shapes of the curves. Similar to the elastic behaviour, the directional dependence of strength is rather insignificant for the KV structures. The minimum peak value is predicted for the [111] orientation, the maximum for the [011] orientation. Note that this is in contrast to the present elastic predictions, where [111] yields the highest Young's modulus and [011] has an intermediate value.

2.3.1 Influence of Structural Disorder

Figure 6 shows the predicted deformations of the SC models with structural disorder of different degree (together with the regular structures) under uniaxial compression for the selected lattice orientations. For small levels of disorder the

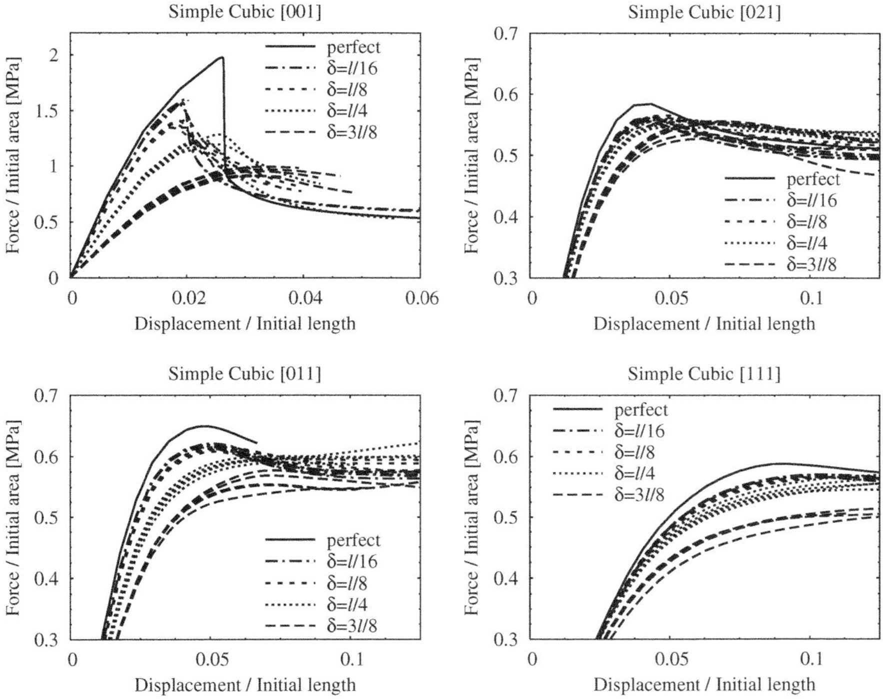

Fig. 7 Predicted overall stress–strain curves of Simple Cubic finite structures with different levels of disorder, δ, for lattice orientations [001], [021], [011], and [111] with respect to uniaxial compression [21]

behaviour is similar to the regular structures and the deformation localizations occur in the same manner. With increasing disorder the localizations are less obvious and at pronounced disorder no localization can be seen in the deformed structures of all orientations.

Figure 7 shows the corresponding stress–strain curves for these cases. The predictions for all five realizations of each single case are shown. Inspection of the [001] orientation reveals, at first, the previous result that for increasing disorder the stiffness of the structures decreases. With increasing deviation of the vertices from their location in the regular structure the governing deformation mechanism changes from pure axial compression in the struts to a mixed mode with increasing bending contribution. As for the stiffness, the peak load is reduced and shifted to higher strains for increased disorder. Also the load drop following the maximum load is less pronounced, resulting in a higher load carrying capacity at greater displacements. For the [021], [011], and [111] orientations the same trends can be seen, but in a much less pronounced way. The structures tend to exhibit more compliant behaviour and lower strength with increasing disorder, and the load drop after the peak is less pronounced. Note that the linear elastic behaviour as discussed earlier, here, is superimposed by nonlinear effects.

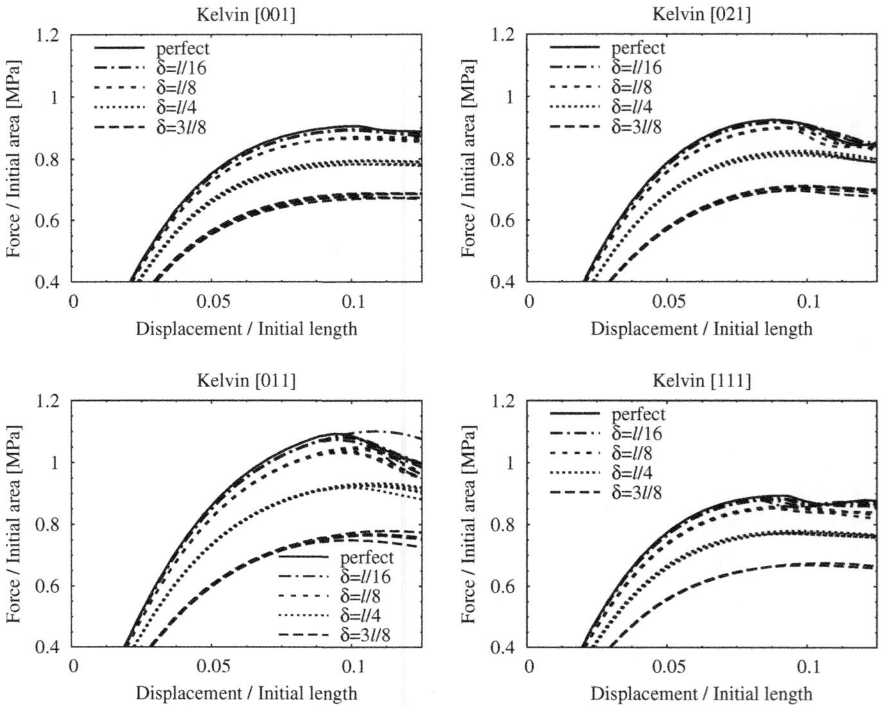

Fig. 8 Predicted overall stress–strain curves of Kelvin finite structures with different levels of disorder, δ, for lattice orientations [001], [021], [011], and [111] with respect to uniaxial compression [21]

For the KV structures the predicted overall stress–strain diagrams for various levels of disorder are given in Fig. 8. For all orientations the introduced disorder leads to a moderate decrease of the peak loads. The qualitative behaviour is not affected by the disorder. These results support the previous finding that KV structures, in general, are not susceptible to deformation localization.

2.3.2 Total Energy Distribution

In the following the spatial energy distribution at the level of individual Finite Elements within regular and disordered SC structures is discussed by means of energy histograms. When a cellular structure is loaded beyond its elastic limit, damage will occur to the structure. Biological materials like bone are able to regenerate themselves, as long as the damage of the tissue is small enough and the integrity of the overall structure is given. In case of error loads beyond the elastic limit, a cellular structure should therefore distribute plastic deformation energy over as many load bearing elements as possible. If localisation occurs and large fractions of the deformation energy are absorbed by few neighbouring elements, localisation of deformation will occur and the structure is prone to fail.

Using the methods described above, it is possible to numerically measure the total energy of each beam element in relation to the mechanical energy that is brought into the sample by externally acting forces. In Fig. 9 the beam elements of a SC-structure with small structural disorder ($\delta = l/16$) are grouped by the deformation energy contained in each beam element. The z-axis (p) denotes the fraction of elements containing a certain total energy at a given moment of the experiment. The x-axis $U_{\text{element}}/U_{\text{overall}}$ refers to the total energy normalized by the overall work. the y-axis denotes the overall compressive displacement Δ. The compression test starts at $\Delta = 0$ and the initial elastic energy is evenly distributed over all elements. Thus there are only beams around $U_{\text{element}}/U_{\text{overall}} = 0$, indicating that many beams are loaded well below the elastic limit and no beams are overloaded.

When the structure is further deformed, certain elements start taking up significantly more energy at $\Delta \approx 0.6$, indicating the beginning of localization of deformation. In the inset of Fig. 9 it can be seen, that one layer in the structure completely collapses and the structure fails since a large number of neighbouring struts is loaded beyond the elastic limit.

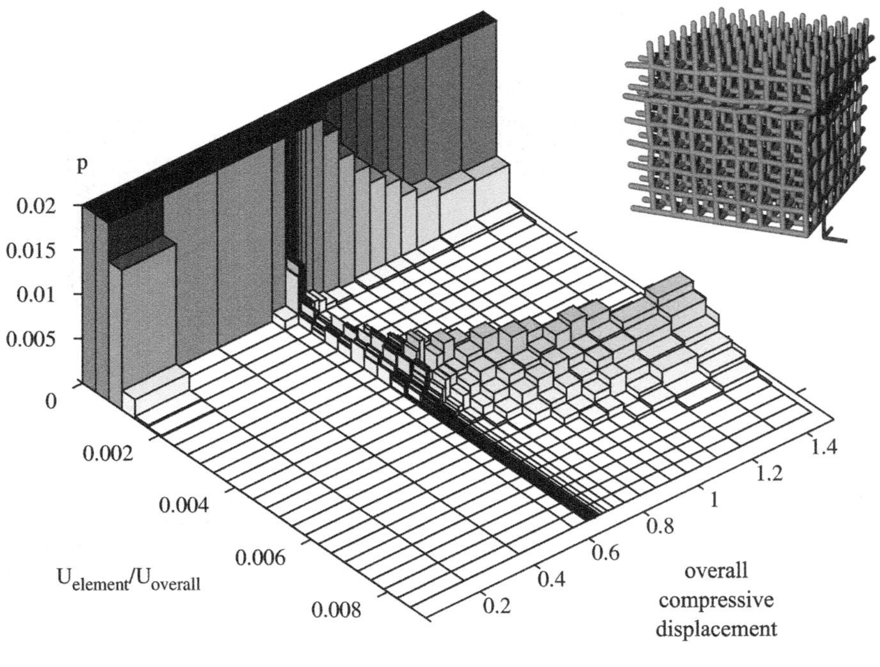

Fig. 9 Predicted energy distribution in the course of uniaxial compression of a disordered ($\delta = l/16$) Simple Cubic finite structure with lattice orientation [001]; total energy within each Finite Element normalized by the overall work, $U_{\text{element}}/U_{\text{overall}}$, vs. fraction of elements, p, (cut off at $p = 0.02$)

Introduction of a larger structural disorder ($\delta = 3l/8$) results in a very different behaviour (Fig. 10): Up to a deformation of $\Delta = 0.6$ the histogram looks comparable to the previous case. But when the deformation further increases, no distinct secondary peak appears in the histogram. Instead, the shoulder of the main peak at $U_{\text{element}}/U_{\text{overall}} = 0$ widens, indicating that the total energy in several elements is increasing. Since the increase in energy uptake is moderate, no element is loaded significantly beyond the elastic limit and localization of deformation cannot be observed. Although above results are only valid for the SC structure loaded in principal directions, similar behaviour can be observed in related cases.

The implications for biological materials are obvious: The peak performance (maximum load or stiffness) in a disordered structure is reduced compared to the perfectly ordered structure. But in case of error loads, the disordered structure is only slightly damaged whereas the ordered structure fails catastrophically due to deformation localization. Since biological materials (e.g. bone) have an ability to regenerate themselves to a certain degree, the local damages caused by error loads can be repaired, as long as the overall integrity of the structure is not lost. This

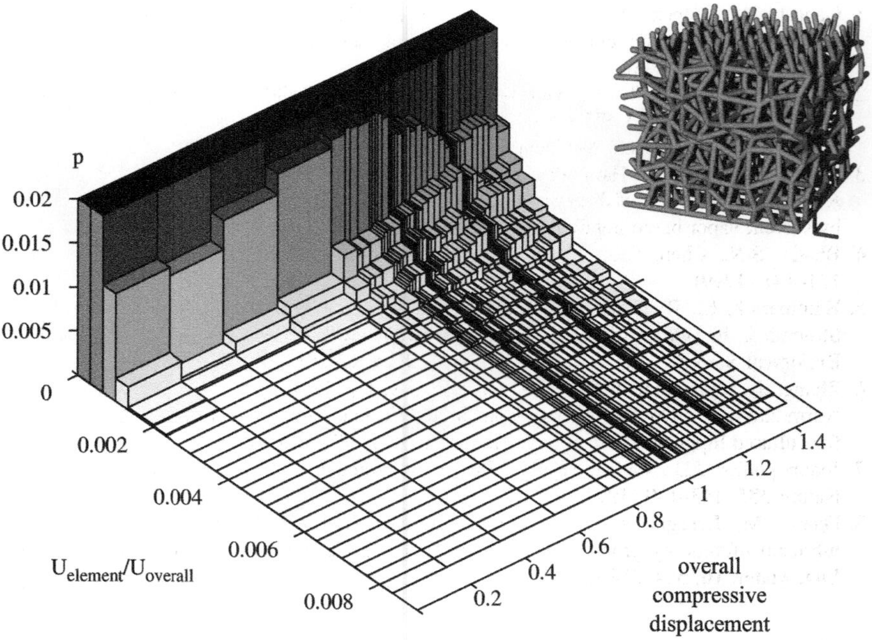

Fig. 10 Predicted energy distribution in the course of uniaxial compression of a disordered ($\delta = 3l/8$) Simple Cubic finite structure with lattice orientation [001]; total energy within each Finite Element normalized by the overall work, $U_{\text{element}}/U_{\text{overall}}$, vs. fraction of elements, p, (cut off at $p = 0.02$)

indicates that, from a mechanical point of view, biological tissues can benefit from the irregularities which are observed in many natural materials.

3 Fabrication of Cellular Structures

Besides lithography-based systems, which are described in the following, several additive manufacturing processes have gained a lot of attention and are available on a commercial basis. Fused deposition modelling (FDM) [23], selective sintering (SLS) [24] and 3D-printing [6, 25] have to be mentioned due to their large installation base. Most of these processes are capable of fabricating cellular structures. Nevertheless lithography-based additive manufacturing systems offer several advantages: On the one hand the feature resolution is significantly better than with all other available technologies. On the other hand the functional properties of the obtained photopolymer can be tuned quite easily by varying the crosslink density of the polymer and by changing the utilized reactive diluents and crosslinkers.

3.1 Laser-Based Stereolithography

With a market share of 42% in the field of service bureaus, laser-based stereolithography is still dominating the high-end market of AMT. The basic working principle of stereolithography (SLA) has not changed since its invention in the 80ies of the twentieth century. The light source of the system is made up of a UV-laser, typically a solid state Nd-YAG-laser. Using a galvanoscanner, the light beam is deflected according to the geometry of the CAD-file. Where the laser beam hits the resin surface, polymerization takes place and the liquid resin solidifies. After one layer has been completed, the part is lowered into the vat and the coating mechanism deposits a fresh layer of resin on top of the part. The process is repeated until all layers have been built.

If the part geometry contains severe undercuts, support structures are necessary since the liquid resin is not capable of stabilizing geometries which have no connection to the main part. The generation of support structures is achieved automatically by modern AMT-software. Problems might arise when the supports, which have to be removed mechanically, have to be detached from delicate features.

Commercial SLA systems typically use layer-thicknesses of 50–100 μm. Due to the high scan-speeds of 500–1,000 mm/s and the large possible build-volumes SLA is mainly used for high-throughput jobs in service bureaus. With an appropriate optical setup the feature resolution and minimal layer thickness can be improved significantly and using micro-SLA-systems lateral resolutions of down to 5 μm have been achieved [26]. Sample structures fabricated by this process are depicted in Fig. 11.

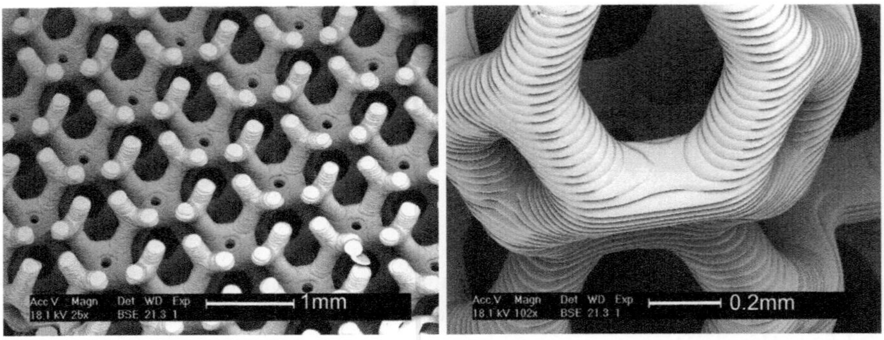

Fig. 11 Cellular structures fabricated by micro-stereolithography [26]

3.2 Dynamic Mask Based Stereolithography

In order to overcome the price-limits given by the fairly expensive UV-laser-systems, lithography-based processes using visible light in combination with a dynamic mask have been developed. The main benefit of these systems is the fact that they can rely on components which are widely used in consumer products (e.g. video beamers).

Dynamic mask based stereolithography (DMS) systems offer distinct benefits and drawbacks compared to laser-based stereolithography: Due to the fact that the whole layer is exposed in one step, the fabrication speed is faster. Typical exposure times for one layer range from 3–12 s. Since most DMS systems use a high-pressure mercury lamp instead of a UV-laser the cost for replacing and maintaining the light source are fairly moderate. The use of high-pressure mercury lamps and micro-mirror arrays necessitates the use of visible light for the polymerization of the resin. Most micro-mirror arrays cannot be used in combination with UV, although recent developments might change this restriction. The light intensity of the available light sources is smaller than in the case of UV-lasers. Both limitations lead to the fact that well established and widely used SLA-resins cannot be used in combination with DMS.

A further drawback is related to the number of pixels which are available on the chip. At a given build-volume the achievable pixel-resolution is given by the number of mirrors on the chip. For a system with $1,400 \times 1,050$ pixels and a targeted resolution of $50\,\mu$m, the build size is 70×50 mm. DMS systems are therefore mostly used for the fabrication of delicate and small parts. If the required build size is very large, the possible resolution will be too low.

A typical schematic setup for a DMS system is given in Fig. 12. The light source emits light which is selectively deflected by the micro-mirrors on the DLP-chip (digital light processing). An appropriate lens focuses the image on the bottom surface of the transparent vat. White pixels lead to a solidification of the photosensitive resin. In regions of black pixels the resin stays liquid. After solidification has taken place, the build platform is lifted in z-direction, according

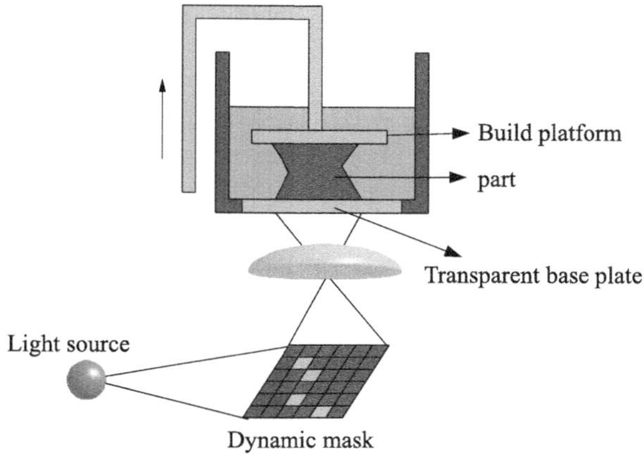

Fig. 12 Working principle of dynamic mask based stereolithography

to the layer thickness. Liquid resin flows into the now empty space and the process is repeated. Exposure from underneath offers several advantages compared to the exposure from top (which is used in traditional laser-based processes). The main advantage from an economical point of view is the fact that significantly less resin is needed to run the bottom-exposure process compared to the top-exposure process. In Fig. 12 the necessary resin level is only a couple of millimeters in height, independently of the height of the part. In a top-exposure setup the resin level has to be high enough, as to accommodate the complete final part.

The main advantage from an engineering point of view is the possibility to expose very thin layers with little processing effort. The layer thickness can easily be adjusted by defining the gap between the previous layer and the transparent bottom of the vat. No special coating mechanism is necessary, and even small layer thicknesses down to 10–20 μm are possible.

One drawback of the exposure from the bottom are issues with adhesion between polymerized resin and the transparent vat. The vat has to be coated with silicone and/or a transparent teflon layer in order to prevent large pull-off forces when the polymerized body is detached from the vat. If the resin contains initiator molecules or monomers with low molecular weight, these compounds tend to diffuse into the vat, leading to unwanted polymerization inside the separation layer.

Due to the fact that the whole layer is exposed in one step, DMS systems work fairly fast. Typically, vertical build speeds between 10–20 mm/h are achieved. Due to the moderate hardware cost, the ability to fabricate complex parts with excellent feature resolution, DMS systems are widely used for applications in jewellery (master patterns for investment casting) and the fabrication of hearing aids.

3.3 Two-Photon Polymerization (2PP)

Two photon polymerization (2PP) [27, 28] offers two distinct benefits with respect to other AMT: (1) The achievable feature resolution is about one order of magnitude better than with other additive manufacturing methods. The minimum achievable wall thickness is currently around 100 nm [29]. (2) Furthermore it is possible to directly write inside a given volume. In contrast, all other AMTs work by shaping individual 2D-layers and stacking up these layers in order to fabricate a 3D-model. Due to this additive stacking-process, it is not possible to embed existing components into a part made by traditional AMT. In contrast, 2PP is capable of writing "around" pre-embedded components. Despite these distinct advantages of 2PP, there are currently no commercial 2PP-applications, mainly due to the low writing speed of 2PP-systems and due to the complexity of the required lasers. With the development of more efficient initiators [30] and more powerful femtosecond lasers, this situation is expected to improve in the near future.

3.3.1 Light Sources and Optical Setup

In order to trigger the non-linear two-photon absorption, light sources with very high photon density are required. Ultra-short pulsed lasers are capable of providing the required intensities. At the same time the average power of ultra-short pulse lasers can be kept reasonably small, since the intensity between the laser pulses is zero. Most currently used 2PP setups therefore rely on femtosecond-lasers with pulse durations between 50 and 150 fs.

The most versatile type of fs-lasers are amplified systems which allow the tuning of wavelength, pulse duration and intensity quite easily. Amplified systems are limited regarding the maximum repetition rate, which is typically in the order of several kHz. The low repetition rate limits the maximum writing speed, since every voxel needs at least one laser pulse to trigger the polymerization. Therefore non-amplified are currently used for most 2PP applications, either based on Ti:Sapphire lasers or on fibre lasers. The utilized laser powers vary between 10 and 700 mW, mostly with pulse durations around 100 fs and repetition rates of 10 to 100 MHz.

A typical setup for a two-photon-lithography system is depicted in Fig. 13. The laser beam first passes through a collimator positioned after the laser head. An acousto-optical modulator (AOM) can split the laser beam into a zero and first order. The first order can be turned on and off by switching the AOM. Then the first order of the collimated beam passes through a $\lambda/2$ wave- plate which can be rotated to adjust the laser power passing the polarized beam splitter. After passing the microscope objective, the laser light penetrates the sample holder containing the photopolymerizable formulation. By guiding the focus of the laser beam through the liquid or solid resin, the three-dimensional structure is polymerized. A camera is positioned behind a semi-transparent mirror in order to allow online-observation of the polymerization process. By illuminating the sample with appropriate lighting (e.g. red light emitting diodes) an improved observation of the sample is possible.

The positioning of the laser focus for the selective photopolymerization is typically achieved by two different methods: (1) The positioning of the laser beam

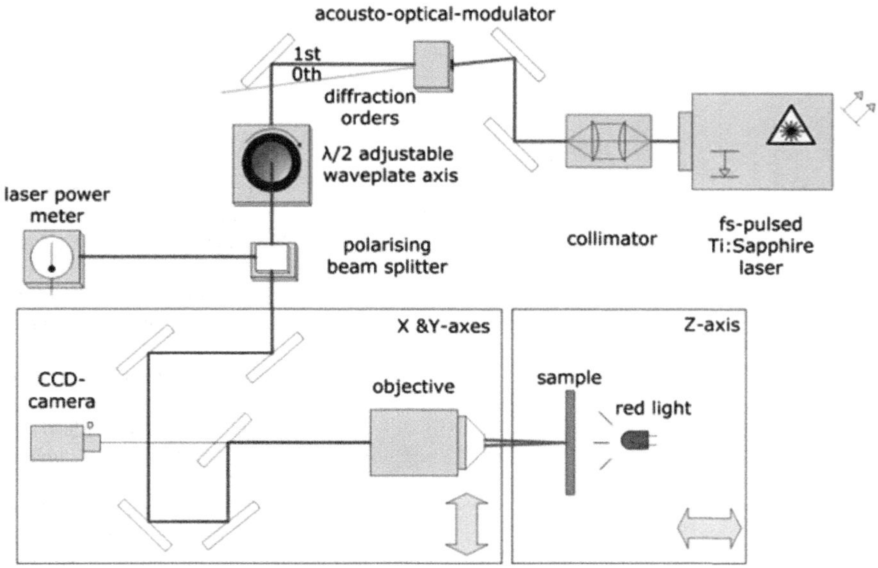

Fig. 13 Schematical setup of a system for two photon lithography

in the xy-plane is done using a piezo-actuated or linear air-bearing stages. Alternatively (2) the laser beam passes a galvanoscanner which is positioned in before the microscope objective. Galvanoscanners have the advantage that laser beam can be positioned precisely and a highly dynamic movement of the beam is possible. The drawbacks are mostly related to the limited build size: For high resolution structures immersion-oil objectives with high magnification (typically $100\times$) have to be used. In combination with a galvanoscanner this setup is limited to build sizes of approximately $30 \times 30\,\mu$m. Piezo-actuated stages have slightly larger scan areas (around $200 \times 200\,\mu$m), with high-precision air-bearing stages significantly larger build areas (up to 100×100 mm) can be covered. Scan speeds up to 30 mm/s are possible when using highly reactive resins and suitable photoinitiators. For parts which require very high resolution and precision, $100\,\mu$m/s up to 1 mm/s are commonly used writing speeds.

3.3.2 Applications

Besides the fabrication of high-resolution structures (see cellular structures in Fig. 14) 2PP is especially suitable for applications where existing components have to be embedded inside the written structure. For instance, 2PP can be used to connect opto-electronic components (laser diodes and photo diodes) by an optical waveguide [31]. Embedding existing components can also be used for applications in tissue engineering: Instead of fabricating a scaffold ex-situ and implanting or

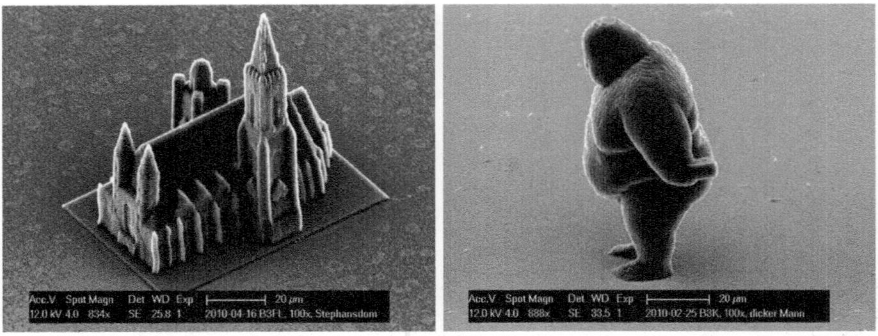

Fig. 14 Structures made by two-photon polymerization

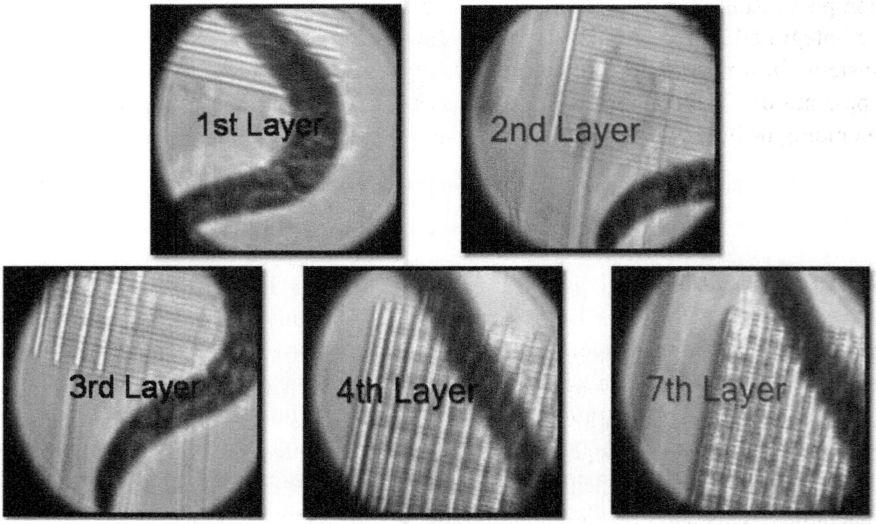

Fig. 15 In vivo writing of scaffold with embedded round worm *C. elegans*, from [32]

cell-seeding the scaffold afterwards, 2PP offers the opportunity to write scaffolds in vivo instead [32].

Living tissue is transparent for infrared light, and thus the 800 nm fs-laser penetrates deep into tissue without causing damage. When embedding the living organism in a 2PP-sensitive resin, a scaffold can be directly written inside or around living cells. The image sequence in Fig. 15 shows an example of a roundworm *C. elegans* which is embedded in an methacrylate-based photopolymer. Using 2PP, a scaffold is written around the worm in several layers, while the organism is still alive.

Due to the high resolution of 2PP, it is possible to write cellular structures with cell sizes in the range of visible light. This is especially intriguing for the fabrication of photonic crystals [33, 34]. 2PP offers a method to systematically study

Fig. 16 Photonic crystals fabricated with 2PP, from [33]

the influence of the cellular architecture on the optical properties of such structures. The conventionally used organic resins are only of limited usability for photonic crystals, since materials with a high dielectric constant are required for tuning the optical properties. Therefore resins with high amounts of inorganic material (e.g. zirconia) have been developed for this application.

References

1. Aizenberg, J., Thanawala, M., Sundar, V., Weaver, J., Morse, D., Fratzl, P.: Materials science: skeleton of *Euplectella* sp.: structural hierarchy from the nanoscale to the macroscale. Science **309**, 275–278 (2005)
2. Gibson, L.J., Ashby, M.F.: Cellular Solids, 2nd edn. Cambridge University Press, Cambridge (1997)

3. Fratzl, P., Weinkamer, R.: Nature's hierarchical materials. Prog. Mater. Sci. **52**(8), 1263–1334 (2007)
4. Meyers, M.A., Chen, P.Y., Lin, A.Y.-M., Seki, Y.: Biological materials: structure and mechanical properties. Prog. Mater. Sci. **53**(1), 1–206 (2008)
5. Lam, C.X.F., Mo, X.M., Teoh, S.H., Hutmacher, D.W.: Scaffold development using 3D printing with starch powder. Mater. Sci. Eng. C **20**, 49–56 (2002)
6. Leukers, B., Gulkan, H., Irsen, S., Milz, S., Tille, C., Schieker, M., Seitz, H.: Hydroxyapatite scaffolds for bone tissue engineering made by 3D printing. J. Mater. Sci. Mater. Med. **16**(12), 1121–1124 (2005)
7. Hutmacher, D.W.: Scaffold design and fabrication technologies for engineering tissues-state-of-the-art and future perspectives. J. Biomater. Sci. Polym. Edn. **12**(1), 107–124 (2001)
8. Bibb, R., Sisias, G.: Bone structure models using stereolithography: a technical note. Rapid Prototyp. J. **8**(1), 25–29 (2002)
9. Woesz, A., Stampfl, J., Fratzl, P.: Cellular solids beyond the apparent density an experimental assessment of mechanical properties. Adv. Eng. Mater. **6**(3), 134–138 (2004)
10. Manjubala, I., Woesz, A., Pilz, C., Rumpler, M., Fratzl-Zelman, N., Roschger, P., Stampfl, J., Fratzl, P.: Biomimetic mineral-organic composite scaffolds with controlled internal architecture. J. Mater. Sci. Mater. Med. **16**, 1111–1119 (2005)
11. Woesz, A., Rumpler, M., Stampfl, J., Varga, F., Fratzl-Zelman, N., Roschger, P., Klaushofer, K., Fratzl, P.: Towards bone replacement materials from calcium phosphates via rapid prototyping and ceramic gelcasting. Mater. Sci. Eng. C **25**(2), 181–186 (2005)
12. Luxner, M.H., Stampfl, J., Pettermann, H.: Numerical simulations of 3D open cell structures—influence of structural irregularities on elasto-plasticity and deformation localization. Int. J. Solids Struct. **44**, 2990–3003 (2007)
13. Luxner, M.H.: Modeling and simulation of highly porous open cell structures—elasto-plasticity and localization versus disorder and defects. PhD thesis, Technische Universität Wien (2006)
14. Silva, M.J., Gibson, L.J.: The effects of non-periodic microstructure and defects on the compressive strength of two-dimensional cellular solids. Int. J. Mech. Sci. **39**(5), 549–563 (1997)
15. Guo, X.E., Gibson, L.J.: Behavior of intact and damaged honeycombs: a finite element study. Int. J. Mech. Sci. **41**(1), 85–105 (1999)
16. Shulmeister, V., Van der Burg, M.W.D., Van der Giessen, E., Marissen, R.: A numerical study of large deformations of low-density elastomeric open-cell foams. Mech. Mater. **30**(2), 125–140 (1998)
17. Li, K., Gao, X.L., Subhash, G.: Effects of cell shape and cell wall thickness variations on the elastic properties of two-dimensional cellular solids. Int. J. Solids Struct. **42**((5–6)), 1777–1795 (2005)
18. Roberts, A.P., Garboczi, E.J.: Elastic properties of model random three-dimensional open-cell solids. J. Mech. Phys. Solids **50**(1), 33–55 (2002)
19. Zhu, H.X., Hobdell, J.R., Windle, A.H.: Effects of cell irregularity on the elastic properties of open-cell foams. Acta Mater. **48**(20), 4893–4900 (2000)
20. Gan, Y.X., Chen, C., Shen, Y.P.: Three-dimensional modeling of the mechanical property of linearly elastic open cell foams. Int. J. Solids Struct. **42**(26), 6628–6642 (2005)
21. Luxner, M.H., Stampfl, J., Pettermann, H.: Finite element modeling concepts and linear analyses of 3D regular open cell structures. J. Mater. Sci. **40**, 5859–5866 (2005)
22. Nye, J.F.: Physical Properties of Crystals. Oxford University Press, Oxford (1985)
23. Zein, I., Hutmacher, D.W., Tan, K.C., Teoh, S.H.: Fused deposition modeling of novel scaffold architectures for tissue engineering applications. Biomaterials **23**(4), 1169–1185 (2002)
24. Agarwala, M., Bourell, D., Beaman, J., Marcus, H., Barlow, J.: Direct selective laser sintering of metals. Rapid Prototyp. J. **1**(1), 26–36 (1995)
25. Moon, J., Grau, J.E., Knezevic, V., Cima, M.J., Sachs, E.M.: Ink-jet printing of binders for ceramic components. J. Am. Ceram. Soc. **85**(4), 755–762 (2002)

26. Stampfl, J., Baudis, S., Heller, C., Liska, R., Neumeister, A., Kling, R., Ostendorf, A., Spitzbart, M.: Photopolymers with tunable mechanical properties processed by laser-based high-resolution stereolithography. J. Micromech. Microeng. **18**((12)), 125014 (2008)
27. Passinger, S., Saifullah, M.S.M., Reinhardt, C., Subramanian, K.R.V., Chichkov, B.N., Welland, M.E.: Direct 3D patterning of TiO2 using femtosecond laser pulses. Adv. Mater. **19**(9), 1218–1221 (2007)
28. Maruo, S., Nakamura, O., Kawata, S.: Three-dimensional microfabrication with two-photon-absorbed photopolymerization. Opt. Lett. **22**(2), 132–134 (1997)
29. Park, S.H., Lim, T.W., Yang, D.Y., Kim, R.H., Lee, K.S.: Improvement of spatial resolution in nano-stereolithography using radical quencher. Macromol. Res. **14**(5), 559–564 (2006)
30. Heller, C., Pucher, N., Seidl, B., Kuna, L., Satzinger, V., Schmidt, V., Lichtenegger, H., Stampfl, J., Liska, R.: One- and two-photon activity of cross-conjugated photoinitiators with bathochromic shift. J. Polym. Sci. A Polym. Chem. **45**, 3280–3291 (2007)
31. Stampfl, J., Infuehr, R., Krivec, S., Liska, R., Lichtenegger, H., Satzinger, V., Schmidt, V., Matsko, N., Grogger, W.: 3D-structuring of optical waveguides with two photon polymerization. In: Kuebler, S.M., Milam, V.T. (eds.) Material Systems and Processes for Three-Dimensional Micro- and Nanoscale Fabrication and Lithography, volume 1179E of Mater. Res. Soc. Symp. Proc., pages 1179–BB01–07, Warrendale, PA, (2009)
32. Torgersen, J., Baudrimont, A., Pucher, N., Stadlmann, K., Cicha, K., Heller, C., Liska, R., Stampfl, J.: In vivo writing using two-photon-polymerization. In: Sugioka, K. (ed.) Proceedings of LPM2010—The 11th International Symposium on Laser Precision Microfabrication, Paper No. 10-42, June 4–7 2010
33. Ovsianikov, A., Shizhou, X., Farsari, M., Vamvakaki, M., Fotakis, C., Chichkov, B.N.: Shrinkage of microstructures produced by two-photon polymerization of Zr-based hybrid photosensitive materials. Opt. Exp. **17**(4), 2143–2148 (2009)
34. von Freymann, G., Chan, T., John, S., Kitaev, V., Ozin, G.A., Deubel, M., Wegener, M.: Subnanometer precision modification of the optical properties of three-dimensional polymer-based photonic crystals. Photon. Nanostruct. **2**, 191–198 (2004)

X-ray Microtomography: Characterisation of Structures and Defect Analysis

Bernhard Harrer and Johann Kastner

Abstract Determination of the three-dimensional (3D) distribution of heterogeneities and structures is of primary concern in the field of materials characterisation and quality control. Heterogeneities may create exploitable property profiles, but they can also degrade reliability and are therefore classified as defects. The quantitative description of heterogeneities is a prerequisite for evaluating their effects and potential for strengthening or degrading a material. Knowledge of the size, shape, location and arrangement of heterogeneities enables the evaluation of the quality of materials and work pieces. Micro-focus X-ray computed tomography (XCT) with flat-panel matrix detectors is the current method of measuring variously absorbing inner or hidden structures without destroying the object. The size and topology of different types of heterogeneities can vary greatly. The different size scales of the heterogeneities and of the affected material volume require appropriate tomographic methods and resolutions. XCT provides statistically significant estimates of volume fractions of heterogeneities in materials depending on the spatial and contrast resolution. The characterisation and quantification of heterogeneities in polymeric materials, light metals, Fe-based materials, metallic foams and metal matrix composites are presented by means of cone beam XCT. Micro-focus and sub-micro-focus X-ray tubes may record intensities related to a minimum volume (voxel size) of $(3\ \mu m)^3$ and $(0.4\ \mu m)^3$, respectively. Spatial resolution also depends on the contrast of the heterogeneity and is usually reliable for features > 20 voxels. Suitable evaluation routines are introduced and rules for

B. Harrer (✉) · J. Kastner
Upper Austria University of Applied Sciences, Stelzhamerstrasse 23, 4600 Wels, Austria
e-mail: bernhard.harrer@fh-wels.at

B. Harrer
Vienna University of Technology, Institute of Materials Science and Technology, Karlsplatz 13, 1040 Vienna, Austria

F. A. Lasagni and A. F. Lasagni (eds.), *Fabrication and Characterization in the Micro-Nano Range*, Advanced Structured Materials, 10,
DOI: 10.1007/978-3-642-17782-8_6,

the detectability and classification of contrast and shape are specified. The architecture of the reinforcement in composite materials is presented and quantified in terms of size distribution, orientation and connectivity. Alignments and the formation of the dendritic structure in cast metals are shown in 3D. A method for differentiation of various heterogeneities which are simultaneously present in one material system is presented. The microstructural features are verified using target metallographic techniques. Thus, important 3D structural information can be achieved to understand their correlation with processing.

1 Introduction

1.1 History of X-ray Computed Tomography

The invention of X-ray computed tomography (CT) is considered to be the greatest innovation in the field of medical diagnostic tools since the discovery of X-rays. The initial use of computed tomography during the 1970s sparked a revolution in the field of medical science because of the enormous benefit of non-invasive imaging of the human body. This technology spread rapidly and in 1980 there were already about 10,000 CT-installations for medical purposes worldwide. Ever since, new developments have led to faster scanning, better dosage and improved image quality. Medical X-ray tomography is well described in literature, e.g. see Hsieh [1] and Buzug [2].

In contrast to the field of medicine, CT for industrial applications was not established as quickly. One of the first users outside the field of medical diagnostics in Europe was the Messerschmidt-Boelkow-Blohm GmbH (MBB, now Eurocopter) company in Munich. MBB has successfully used conventional medical CT-scanners for non-destructive testing (NDT) of technical parts since 1979, in particular for testing helicopter components like rotor blades [3].

The Federal Institute for Materials Research and Testing (BAM) in Berlin started the development of a CT-system for technical applications in 1978. The start-up of the system, equipped with a radionuclide source, was in 1983. Among other applications, the device was used to investigate barrels of nuclear waste [4].

In the same period Habermehl and Ridder [5] developed a transportable CT-system, which was used to study the water transportation of trees. Further CT-devices were used for industrial purposes such as an NDT-technique in the late 1980s for analysing ingots, turbine blades or parts of intercontinental missiles.

Until the 1990s, however, industrial CT was only applied in special cases when other testing methods failed, or was considered as a complementary method in the field of non destructive testing. Although the advantage of CT-techniques to image the form, size and position of heterogeneities in three-dimensions (3D) was demonstrated, the spatial resolution was not good enough in the engineering applications to reproduce the triumphant success of medical CT during the 1990s.

Table 1 Historical industrial XCT milestones

Year	Milestone
1895	Wilhelm Conrad Röntgen discovered a new kind of radiation, which he named the X-rays
1917	Johann Radon demonstrated mathematically that an object could be replicated from an infinite set of its projections
1963	Allan MacLeod Cormack determined density distributions of aluminium and plastic by means of X-ray tomography in a laboratory simulation
1967	Godfrey Newbold Hounsfield demonstrated the first clinical CT-scanner
1971	The first patient with a large cyst was scanned and pathology was visible in the image
1979	MBB (Eurocopter) used medical CT-devices for testing helicopter rotor blades
1983	Start-up of the first industrial CT-scanner with a Cobalt-60 radionuclide source developed by BAM
1995	The first XCT-systems equipped with matrix detectors were used for geometrical measurements (e.g. aluminium components)
2004	VDI/VDE Society for Metrology and Automation Engineering are drawing up guidelines for geometrical measurement using CT
2005	Desktop devices with sub-micro-focus tubes and a spatial resolution in the sub-micro range were introduced
2010	> 500 XCT-installations for industrial and scientific applications in Europe. XCT is widely used for metrology, non-destructive testing and materials characterisation

The break-through for industrial CT in Europe occurred at the end of the 1990s when the potential of CT for three-dimensional geometrical measurements was recognised. An important pioneer in this application was the European motor vehicle industry, which applied CT for initial sample testing of engine components made of aluminium castings. Due to the rapid development of high-performance computers and software for processing and visualisation of 3D-datasets, as well as new solid-state flat-panel matrix detectors with a high dynamic range on the basis of amorphous silicon crystals instead of X-ray image amplifiers, new applications were exploited. In particular, the development of the micro-focus-CT with cone beam geometry and comparatively short measurement times increased the acceptance of this technology for industrial applications enormously. Today industrial micro-focus X-ray computed tomography (XCT) is an efficient and highly regarded NDT method that provides entire 100% inspection and accurate 3D volume modelling of almost any object, with the exception of only some objects made of heavy metals. Table 1 gives an overview of important milestones in the development of industrial XCT.

1.2 Generation of X-rays

The conventional method of producing X-rays in a laboratory is to use an evacuated tube with an acceleration voltage of up to 600 kV [6]. The tube contains a cathode filament where an electron cloud is formed and these electrons are subsequently accelerated towards the anode. The anode is a water-cooled target

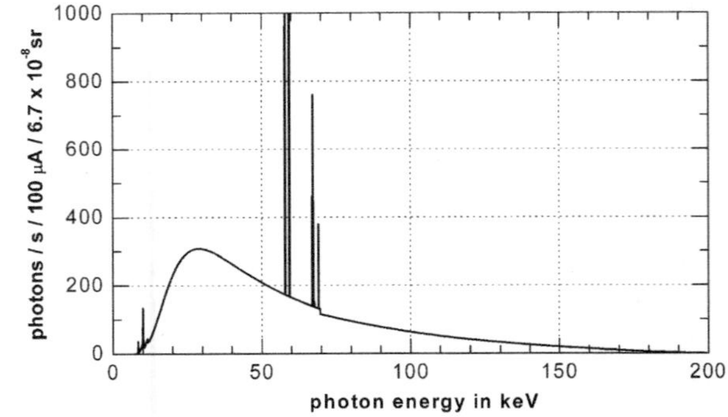

Fig. 1 Simulated X-ray energy spectrum of an X-ray tube with a tungsten target operating at 200 kV (energy channel bandwidth $_{\Delta}E$ is 100 eV) [8]. The spikes correspond to the characteristic radiation of the tungsten target

usually made of tungsten or molybdenum. When the accelerated electrons interact with the target material different types of collisions take place and generate an X-ray spectrum as shown in Fig. 1. The number of photons per second and per 100 μA target current related to the solid angle (steradiant) are plotted against the energy.

As the accelerated electrons reach the target the majority of the interactions involve small energy transfers leading to ionisation of the target's atoms and to heating of the object. For a typical X-ray tube, over 99% of the input energy is converted into heat [1]. For this reason cooling the target, and thus the set-up (direct beam set-up or transmission tube set-up), plays an important role. The more interesting interaction is the so-called 'Bremsstrahlung': X-rays are emitted as the electrons slow down in the target which leads to a continuous spectrum. Some narrow spikes are superimposed on this spectrum at certain energies. The wavelengths of these characteristic X-ray lines were first shown by Moseley [7] to be a function of the atomic number of the target's material.

1.3 Photon-Matter Interaction

The X-rays are known to have a very high, material-dependent capability of matter penetration, due to to their high energy. But X-ray photons are also absorbed and scattered by matter. Several fundamental ways in which X-ray photons interact with matter are known: the most important physical mechanisms are photoelectric absorption, Rayleigh or Thomson scattering, Compton scattering and pair production which are explained by Moore [9] and Eisberg [10].

Most of the mechanisms of absorption involve the conversion of the photon's energy to another form, while some simply change the photon's direction. For each mechanism the probability that photons interact with the material can be defined in a single attenuation coefficient. An attenuation coefficient is a function of the photon energy (E), the mass density (ρ) and the atomic number (Z) of the transmitted material and can be expressed in two ways: on the one hand by means of the

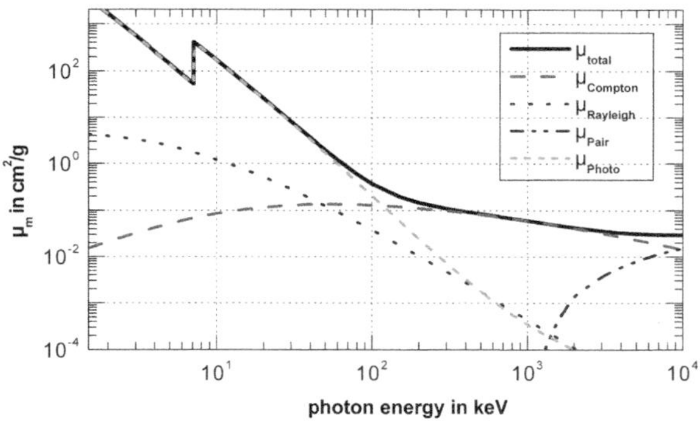

Fig. 2 Total mass attenuation coefficient (μ_{total}) for iron against the photon energy decomposed into the photoelectric absorption (μ_{Photo}), Rayleigh scattering (μ_{Rayleigh}), Compton scattering (μ_{Compton}) and pair production (μ_{Pair}). For X-ray tubes relevant areas are up to 600 keV. The data was taken from the NIST Standard Reference Database [13]

linear attenuation coefficient μ_{l}(E,ρ,Z) in (cm^{-1}) and on the other hand by the mass attenuation coefficient μ_{m}(E,Z) in (cm^2/g). Both coefficients are linked with each other by the mass density. The total probability of interaction is the sum of the probabilities of each physical mechanism which is shown as the bolt line in Fig. 2 which is exemplary for iron. The individual interaction probabilities are strongly dependent on the energy of the incident photon. While for small energies below 100 keV photoelectric absorption is the dominant effect, for energies above 100 keV the total attenuation is almost entirely caused by Compton scattering interaction. With regard to the probability of interactions, Rayleigh scattering is relevant, though not dominant. Pair production, however, has a relevant contribution only for energies above 1 MeV, and can therefore be disregarded for X-ray tubes.

For the quantitative description of absorption and scattering of electromagnetic radiation, $I_0(E)$ is considered as the incident photon flux generated, for example, by an X-ray tube (see Sect. 1.2). These photons interact with the matter of a sample of thickness s and after absorption and scattering, $I(E)$ is ascertained by a detector. The relationship between $I_0(E)$ and $I(E)$ can be expressed with the Lambert–Beers [11, 12] law, or the exponential attenuation law by Eq. 1.

$$I(E) = I_0(E) \cdot e^{-\int_0^s \mu_{l,s}(E,\rho,Z)\cdot ds} \tag{1}$$

By normalising both sides of Eq. 1 with $I_0(E)$ the normalised projection value is obtained. Computation of the logarithm of this fraction gives the line integral of the linear attenuation coefficient μ_{l}(E,ρ,Z) along the X-ray trajectory through the sample. After the application of mathematical reconstruction algorithms each position within the dataset is represented by a local attenuation value.

1.4 Detection of X-rays

The detection of X-ray photons is based on the same principles of interactions between X-ray radiation and material described in Sect. 1.3. The purpose of a detector is to convert a photon flux *I*(*E*) into an electrical signal.

In the first tomographic systems gas detectors were used, well known as Geiger–Müller counters. When an X-ray photon enters the detector, it ionises the gas (e.g. Xenon) and forms ions and electrons. High voltage is applied between a cathode and an anode. Thus the electrons are accelerated towards the anode and this process causes a current which can be measured and is related to the intensity of the radiation [2].

Apart from gas detectors there are numerous different types of digital X-ray detectors, like image intensifiers, flat screen detectors, linear detector arrays and X-ray sensitive CCD cameras. For XCT the most important and common ones are digital flat-panel detectors. These detectors can be divided into two types: direct and indirect conversion detectors. Both types are based on thin layers of amorphous silicon (a-Si) or other semiconductors deposited onto glass substrates, with two-dimensional arrays of detector elements fabricated on the a-Si. For the indirect detection systems, X-ray photons pass onto a phosphorescent screen (usually Gd_2O_2S:Tb or CsI:Tl) and produce visible light, which is in turn detected by the array of thin film transistor (TFT)-switched photodiodes. For detectors with direct conversion, the radiation is directly converted in the sensor layer which is based on Se, CdTe or ZnCdTe [14–18].

Important quality features for digital detectors are the amount and size of the detector elements (pixels), the dynamic range, signal to noise ratio (SNR), contrast noise ratio (CNR) and the detective quantum efficiency (DQE). Quality features for the individual pixels are their homogeneity, noise behaviour, image lag as well as the number, distribution and position of faulty pixels [15].

2 State of the Art

2.1 Principle of Cone Beam CT

Due to the measurement speed and quality, CT-systems with cone beam geometry and flat-panel matrix detectors [17, 18] have gained general acceptance in the market of CT-systems for industrial applications and materials characterisation. Figure 3 shows the principle of cone beam CT using absorption contrast with the main components: X-ray tube, rotating stage and detector. Using CT, a specimen is placed on a precision rotating stage between the X-ray source and a flat-panel detector. The X-ray tube produces a conic beam of photons that transmit the specimen to be analysed. Due to this geometry a magnified projection image (2D radiograph) is recorded on the digital detector. Reducing the source-to-object

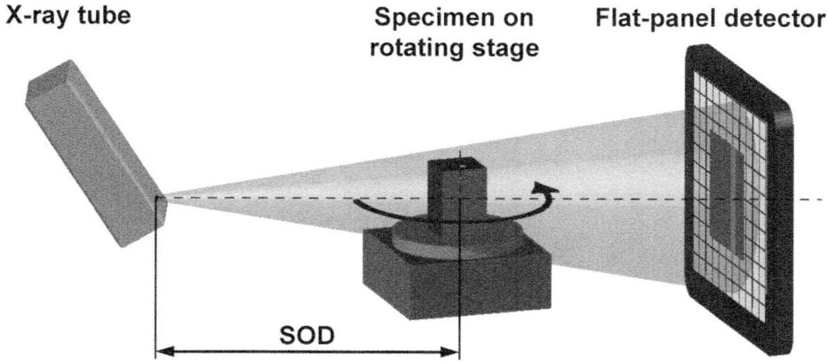

Fig. 3 Principle of XCT with cone beam geometry: main components and variable source-to-object distance (SOD) for varying the magnification (voxel size)

distance (SOD) by moving the rotating stage with the specimen allows enhancement of the magnification and therefore of the spatial resolution, but reduces the field of view.

The specimen is rotated step by step, taking projection images at each angular position. One scan usually covers a rotation of 360° by steps of just a few tenths of a degree up to 1°. For specific applications a limited angle scan [19, 20] can be performed.

Feldkamp et al. [21] describe an approximating reconstruction algorithm from a series of 2D radiographs for circular cone beam tomography. It is an approximation in the sense that the reconstruction result will deviate somewhat from the measured object regardless of the measurement resolution. For moderate cone angles, these differences are however small and often negligible [22, 23]. The simplicity of this method has made it the most used algorithm for cone beam reconstruction. The result of the reconstruction algorithm is a volume dataset which consists of discrete volume elements. A volume element is called a voxel and is a contraction of the words *volume* and *pixel*. Each voxel represents a grey value which corresponds to the effective X-ray attenuation coefficient (see Sect. 1.3) at each position of the dataset.

2.2 CT-Devices

To describe the technical state of CT-devices one has to differentiate between cone beam CT equipped with X-ray tubes and systems which are typically installed at synchrotron facilities using the beamline as a monochromatic source. Compared to X-ray tubes, synchrotron radiation offers a significant advantage by its nearly parallel beam of high brilliance delivering a very high flux (at least 1,000 times larger than X-ray tubes for third generation synchrotrons [24]). The contrast resolution can be increased by applying monochromatic radiation, which also makes it possible to produce phase contrast of internal interfaces between regions not providing sufficient absorption contrast [14]. These advantages mean reduced artefacts and improved resolution in comparison to CT-devices with tubes.

Cone beam XCT covers a wide range of CT-devices from desktop ones to the largest that need a special radiation protection room. Furthermore, there are systems which can be attached in a scanning-electron-microscope (SEM) to allow 3D imaging of small samples with a resolution > 500 nm without compromising standard imaging modes of the SEM [25]. Figure 4 shows two typical commercially available CT-devices which this sub-section focuses on.

The X-ray tubes (open or sealed) used in cone beam CT-devices are characterised by accelerating voltage, maximum tube current, focal spot size and the intensity output. Important for the image quality are the diameter and stability of the focal spot. Generally, the smaller the focal spot size the better, but as the electron flux is focused on a smaller area, the power of the tube must be reduced to prevent overheating at the tube anode. Therefore, the focal spot size becomes a trade-off between resolution capability and power.

According to the Association of the German Automotive Industry, tubes can be classified as normal-focus, mini-focus, micro-focus and sub-micro-focus [15]. Normal-focus tubes (typically up to 450 kV) have focal spots in the range of approximately 0.3–2.0 mm and mini-focus units have focal spots ranging from 0.05 to 0.5 mm. The focal spot size of a micro-focus tube depends on the applied output and is generally 5–50 μm [15]. Micro-focus tubes are typically used for acceleration voltages of up to 250 kV and 300 W maximum output. Due to the high power on the target a direct beam set-up with active cooling is used for these devices. As an example the CT-system *Rayscan 250X* is shown in Fig. 4a. It is equipped with two X-ray sources for the inspection of a wide variety of objects, ranging from micro-parts to cylinder heads. The system is housed in a fully air-conditioned radiation protection area to suppress thermal expansion variations.

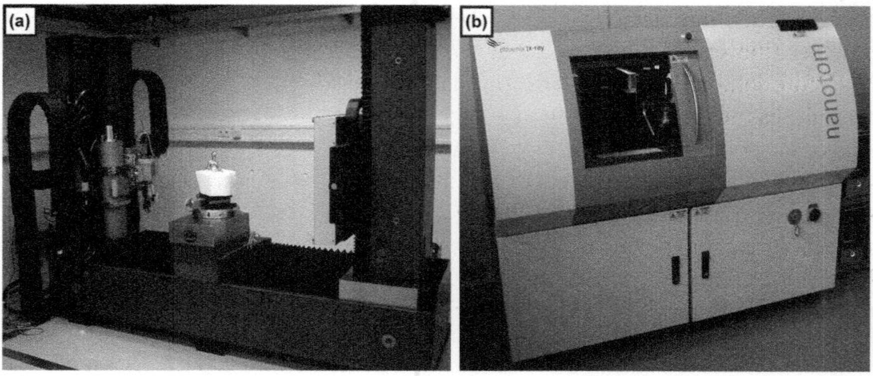

Fig. 4 Commercially available cone beam CT-devices: **a** CT-system *Rayscan 250X* with a 225 kV micro-focus and a 450 kV normal-focus tube (*left*), rotating stage with specimen and flat-panel detector (*right*) mounted on granite columns. **b** Desktop device *nanotom 180NF* with a 180 kV sub-micro-focus tube (transmission target) encased in a radiation safe cabinet

It stands on its own base/foundations in order to be isolated from the rest of the building to eliminate vibrations. The system is equipped with a 225 kV micro-focus and a 450 kV normal-focus tube to add flexibility to the system. The target material for both tubes is tungsten. The *PerkinElmer* detector is a 16-bit solid-state 1,024 × 1,024 pixels^2 a-Si scintillation flat-panel detector with an active area of 409 × 409 mm^2 and a pixel size of 400 × 400 μm^2 [26]. Beside the pixel size of the detector and the size of the focal spot, the resolution quality of the system also depends on the spatial stability of all components. In order to ensure high measuring accuracy and temperature stability, the system uses a high-precision manipulation system made of granite with a precision of about 1–2 μm [27]. Resolution of detail is also strongly influenced by the measurement parameters applied, the size and material of the sample. As a guideline the resolution of detail amounts > 1 μm for the micro-focus tube and > 100 μm for the normal-focus tube [28], assuming a sample size of a few millimetres. The source-to-detector distance is 1.5 m and defines the maximum sample size. Typical scan times are 25–90 min.

Recently, XCTs in the form of a miniaturised design of the above mentioned system have become commercially available. They are produced as desktop devices and have a measurement chamber that is entirely shielded against X-ray beams by lead walls, so that no further means of protection are necessary. Again the achievable resolution or image sharpness is primarily influenced by the focal spot size of the X-ray tube. The tubes used for desktop devices are usually equipped with a transmission target in order to achieve very small spot sizes. The X-ray source is located very close to the outer wall of the X-ray tube allowing the user to bring samples very close to the source to ensure high magnification. However, in this set-up no active cooling of the target is possible. For this reason the anode current cannot be very large which limits the applied maximum output. Such sub-micro-focus tubes typically use up to 180 kV with a maximum output of 15 W. The spot size varies from a few microns to 0.8 μm for the latest transmission target tubes [29]. Figure 4b shows the desktop CT-device *nanotom 180NF* which uses a 180 kV sub-micro-focus tube with an exchangeable transmission target and a 5-megapixel *Hamamatsu* flat panel detector with an active area of 120 × 120 mm^2 (2,316 × 2,316 pixels^2, 50 × 50 μm^2 pixel size) [30]. The source-to-detector distance is about 0.5 m and samples with a maximum diameter of 120 mm can be scanned. Representative scan times are 1.5–4 h. Typical sample sizes are in the range of millimetres or less to achieve a resolution of detail in the range of sub-microns.

2.3 Artefacts and Their Correction

In addition to the limited resolution and geometrical accuracy, the major problems of cone beam CT are measurement artefacts [31]. Artefacts are artificial structures in the resulting dataset which do not correspond to a real sample feature. They are components of image discrepancy that are not explained by noise or by

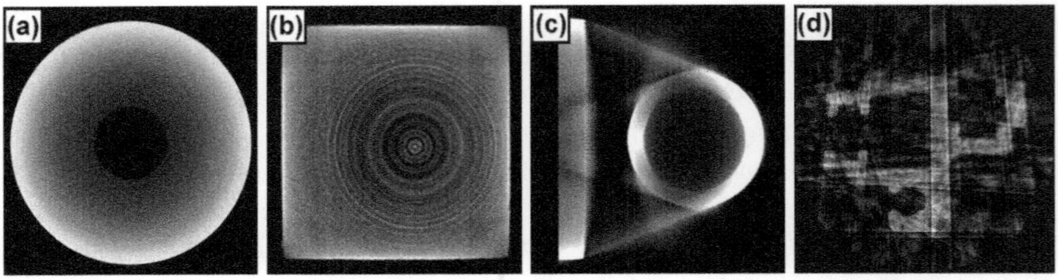

Fig. 5 Typical artefact types in CT: **a** beam hardening artefact (Cupping effect), **b** ring artefacts, **c** streaking artefacts, **d** artefacts due to noncompliance with the sampling theorem

transmission properties and which lead to problems with measurement interpretation and dimensional accuracy.

One of the most common types of artefact are cupping artefacts based on beam hardening, shown in Fig. 5a. These are caused by using a polychromatic X-ray spectrum emitted by the tube and the strong dependence of the attenuation coefficient of the interacting material on the photon energy (see Sects. 1.2 and 1.3). If an object is penetrated, the low-energy radiation of the beam is more strongly absorbed with increasing penetration depth. Only the high-energy radiation penetrates the object and contributes to image generation. Furthermore, the detector efficiency is a function of the photon energy. As a result, these regions are assigned to grey values which are too low in the reconstructed dataset. For materials with a high atomic number and of high density (e.g. Fe, Cu or W) these artefacts distort the image and they can be reduced to a certain degree by filtering the X-ray spectrum with aluminium and copper filters, thus removing low-energy photons before the X-rays enter the object and/or appropriate software corrections.

For mono-material components one common or widely-used correction procedure is linearisation, which was formulated by Herman [32] in 1979. The principle of linearisation consists of transforming the measured polychromatic projection values into corresponding monochromatic values. In practice, a non-linear characteristic correction curve is determined experimentally from projection values of homogeneous reference samples with well-known geometries. This correction curve is applied to the projection images (2D radiographs) as a pre-processing step before reconstruction starts. Several compensation and correction strategies have been discussed in the relevant literature and related or enhanced methods were introduced by Hopkins [33], Hammersberg [34] or Van de Casteele [35]. Kasperl [36] proposed an iterative algorithm to reduce scattered radiation and beam hardening in cone beam CT called Iterative Artefact Reduction (IAR). Projection images are pre-processed using the correction curve which is extracted from a post-processing step of reconstructed data. Therefore the correction curve is enhanced by each iteration and consequently the quality of the dataset is improved. The major disadvantage of all these methods is that for each specimen and material a new characteristic curve has to be determined and one reconstruction per iteration is necessary.

Scattered radiation plays an important role for materials with high attenuation coefficients. Scattering occurs when the incident X-ray photon is deflected from its original path by an interaction with matter (see Sect. 1.3). These scattered X-rays may become a considerable part of the overall signal which results in artefacts. One correction strategy for scattered radiation was introduced by Joseph [37] but this method works only for simple object geometries. Methods which consider the exact geometries (e.g. Raytracing-methods or Monte-Carlo simulations [38]) are computationally intensive and unacceptable for industrial applications to date [36].

Ring artefacts result from (thermal) sensibility drift of the individual detector pixels or faulty pixels. As shown in Fig. 5b, they appear as concentric rings. A possibility for their compensation is a random, horizontal movement of the detector between the acquisition of each projection image.

If highly attenuating objects are present within a sample, streaking artefacts which are shown in Fig. 5c can appear. Furthermore, Fig. 5d illustrates artefacts caused by significantly higher angular projection steps than according to the sampling theorem (i.e. too few projection images were recorded during the scan). Other important artefacts are:

- Partial volume artefacts: an object usually has a sharply contrasted edge. Due to the limited resolution of the detector, this edge will usually not be located directly at the boundary of one detector element to another. Therefore, the intensity of the X-ray on this element will be averaged over the detector width and the object will be blurred. Reducing the voxel size decreases this effect.
- Motion artefacts: are caused by movement of the object, components of the CT-device or the focal spot during the scan.
- Reconstruction artefacts: resulting from the approximation features of the computation operations—to date usually the filtered backprojection (Feldkamp reconstruction algorithm [21]) is applied.

2.4 CT-Data Evaluation Methods

An essential pre-requisite of feature recognition for quantitative analysis is accurate segmentation of the objects of interest in the CT-dataset. As CT-scans are affected by noise and artefacts to a certain degree, image enhancement by applying adequate filters is often necessary before the segmentation process is executed. For this prefiltering/denoising step, e.g. Gaussian-, anisotropic diffusion- or wavelet-filters are applied.

The easiest grey value-based segmentation methods are thresholding methods using either global or local image information. In thresholding, the image segment of interest can be characterised by its brightness: material structures with an attenuation coefficient greater than air will result in bright voxels. The simplest method to binarise an image is to select a grey value with the visual support of a histogram and a preview showing the binarised result on one slice [39]. Apart from visual determination of the appropriate threshold, automatic histogram-based thresholding methods (e.g. Prewitt and Mendelsohn [40] or Otsu [41]) are available. All these techniques have the disadvantage that they rely on grey level information. Such methods can be very useful for initialisation of more sophisticated algorithms such as region-based segmentation. Region-growing starts from seed points (usually provided by the operator) and neighbouring voxels are examined one at a time and added to the growing region if they are sufficiently similar. Comparison may be made to the entire region or only to the local pixels (voxels), with the latter method allowing gradual variations in brightness. The procedure continues until no more pixels (voxels) can be added [42].

In order to generate object features (characteristics like volume, surface area, shape factors, etc.) labelling (object filling, region detection) of connected components is necessary. The prototype for labelling algorithms is the well-known Rosenfeld–Pfaltz method [43].

Segmentation of grey value images into regions for quantification is probably the most important procedure for image analysis. Many novel techniques have been used but they are rather narrow in their range of application.

2.5 Applications

XCT has gained increasing importance and acceptance for the 3D examination and inspection of industrial specimens in recent years. Generally, there are two major application areas of XCT in science and industry:

- Non-destructive testing of materials [9, 39]: materials characterisation using the absorption contrast in XCT is the method of choice for many applications since it is applicable to all materials and the achievable resolutions are adequate for most applications.
- Dimensional measurement (metrology) which is a relatively new field of research [44]. The primary interest in metrology is the measurement of inspection features like crucial distances, wall-thicknesses and diameters. Currently only XCT is able to measure internal or hidden structures completely without destroying the specimen. The data obtained by XCT can also be used to create datasets needed as an input for rapid prototyping machines (e.g. for selective laser sintering).

In quality assurance, metrology is a very common method which measures the surface geometry of a component using tactile or, more recently, optical sensors. XCT is an alternative to co-ordinate measuring machines and in addition not only obtains surface information but also accurate information about internal structures which are not accessible to tactile measuring systems. Furthermore XCT can even be used to characterise partly or fully assembled components.

A common method for the visualisation of geometrical deviations is an actual/nominal comparison with colour coding: a comparison for an Al-diecasting part is

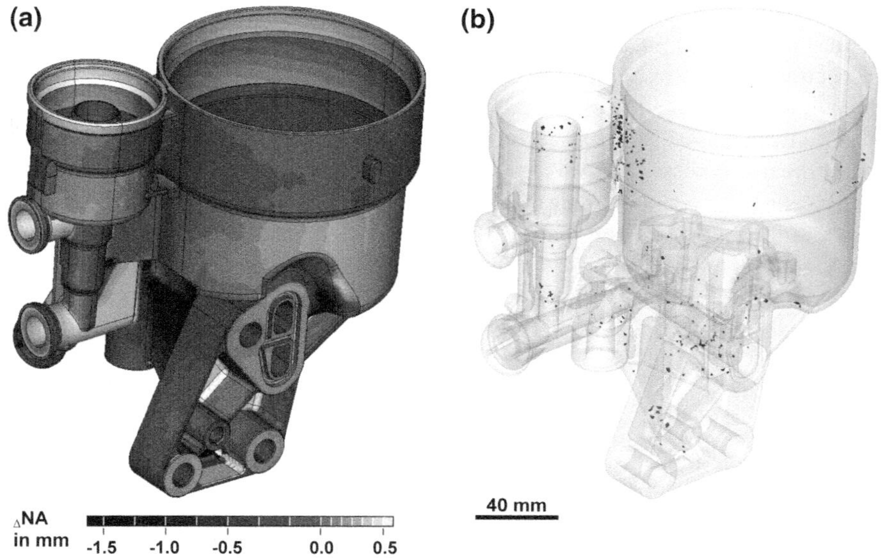

Fig. 6 Main application areas of XCT demonstrated with the help of an Al-diecasting part: **a** metrology: actual/nominal comparison between CT-data and CAD model (colours correspond to the local deviation $_{\Delta}$NA). **b** non-destructive testing: semi-transparent 3D porosity analysis (pores are *coloured black*). The voxel size was (234 μm)^3

shown in Fig. 6a. The colour scale denominates the deviations of the CT-data (isosurface) from the reference (CAD model). Each position of the reference object is coded with a colour corresponding to the local deviation $_{\Delta}$NA: dark areas refer to strong negative and light grey areas to positive deviations.

There are three major problems in the field of metrology: various characteristics of artefacts, the limited resolution and the limited geometrical accuracy. In some cases artefacts even prevent reliable dimensional measurement. Issues such as inappropriate parameter settings for measurement, excessive penetration lengths, or multi-material components are only some of the factors responsible for inaccurate measurement results. Another big challenge for the application of XCT for metrology is the accurate detection of the surface (interface of material and air) or the interface between two different materials. Many local threshold methods are under investigation but no general solution has been found so far. A further critical problem is the lack of information on the uncertainty of an extracted surface or interface, and in consequence, on the quality of the extracted dimensions. For these reasons the application of XCT for dimensional measurement is still very limited, and the problems are even more significant if a component consisting of different materials (like a polymeric plug with metallic pins or joinings between copper and graphite) is to be measured by XCT.

The increasing complexity of innovative products as well as ever more stringent requirements with regard to quality and reliability call for more detailed knowledge about internal structures of materials and components. Hence, non-destructive and contact-free techniques for the 3D-characterisation of heterogeneities will be an increasing demand in process development and production, especially for the inspection of complex geometries and critical features located inside materials. The detection of heterogeneities such as voids, gas pores or cracks in light metals or plastics by means of XCT has become a common non-destructive testing method in recent years. Porosity analysis is performed using CT-data in order to validate manufacturing processes. Figure 6b shows the result of a porosity analysis of an Al-die-casting part in semi-transparent 3D. The detected shrinkage pores are coloured black and agglomerations can be identified. XCT for non-destructive testing has been applied in many fields of materials science [14, 24, 39, 45] and the examples in Sect. 3 below are by no means exhaustive. However, a classification of some of the studies is possible.

3 Characterisation of Heterogeneities

3.1 Polymeric Systems

There are several applications of XCT for characterisation of polymeric systems. The most important are:

- characterization of fibre reinforced polymer—measurement of fibre length distribution and fibre orientation.
- characterization/measurement of porosity.
- characterization of other polymeric systems like particle reinforced polymers or wood plastic composites.

Fibre reinforced polymers (FRP) are competitive with regard to stiffness and strength-to-weight ratio compared to metallic materials. Therefore the applications of these materials have become more and more important in light weight products. Non-destructive and contact-free techniques for the 3D-characterisation of fibre composites are increasingly used in process and production development. XCT is an adequate method to measure the fibre orientation and fibre length distribution for glass fibre and carbon fibre reinforced polymers [46]. Figure 7 shows typical XCT-results. In both pictures a clear contrast between the fibres and the polymeric matrix can be seen.

The CT-data can be processed by various 3D filters such as anisotropic diffusion, then threshold and thinning operations to ascertain the fibre length distribution and fibre orientation. The applied workflow is based on the work of Tan [47]. The anisotropic diffusion filter leads to better contrast between fibres and the polymeric matrix. Therefore, the individual fibres can be extracted by thresholding and the medial axis of every fibre can be determined by a thinning algorithm. From these, data fibre lengths and the three-dimensional orientation can be determined. The final result of the CT-evaluation procedure is a three-dimensional

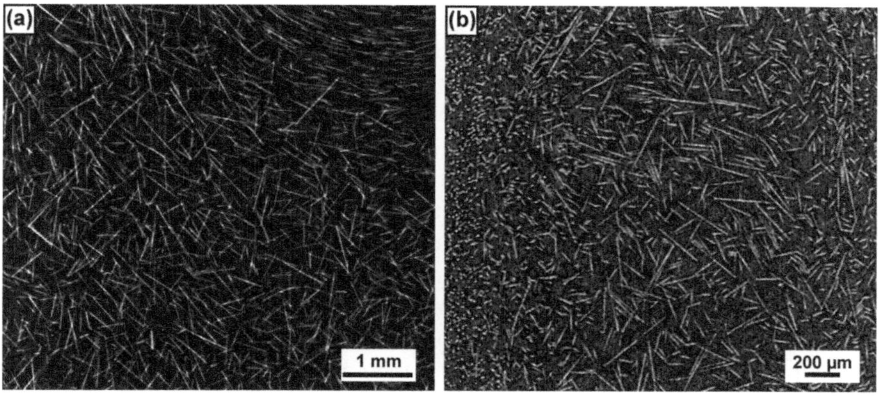

Fig. 7 XCT-tomograms of fibre reinforced polymers (FRP): **a** glass fibres in a polybutylene terephthalate (PBT) matrix, voxel size = (6.0 μm)^3; **b** carbon fibre in a polyamide (PA) matrix, voxel size = (2.0 μm)^3

representation and visualisation of the fibre distribution by means of colour coded vector-fields and glyphs.

The presented approach was tested on fibre reinforced gasket rings [48]. The first sample is an injection moulded gasket ring, polyetheretherketone (PEEK) with 30 vol.% fibre fraction, the second sample was an extruded and turned gasket ring, PEEK, with 30 vol.% fibre fraction. Both samples had a diameter of 9 mm and were scanned with a voxel size of (12 μm)^3. The results of the fibre extraction process are presented in Fig. 8 where dark areas indicate the orientation of the fibres. The fibres are distributed homogenously in both gasket rings. The arrow in

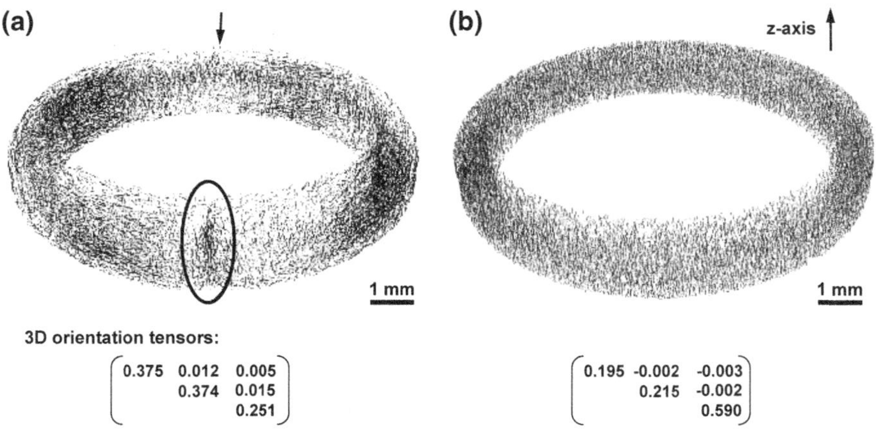

Fig. 8 Visualisation of segmented fibres and specification of the 3D orientation tensors of **a** an injection moulded gasket ring and **b** of an extruded gasket ring. *Dark areas* correspond to the orientation of the fibres. The *arrow* in **a** marks the gate and the *ellipse* marks the weld line of the injection moulded part

Fig. 8a shows the gate, and the ellipse marks the weld line of the injection moulded ring. It can be determined that the fibres within the injection moulded part are mainly orientated parallel to the surface of the ring, whereas in the extruded part the fibres are orientated in the same direction. These two examples demonstrate the usefulness of XCT for extraction of the fibre orientation.

From the CT-data of the two gasket rings the fibre orientation tensor can be determined. The orientation tensors for the two rings are presented in Fig. 8. The matrix diagonal components of the orientation tensors provide information about the texture in orientation. Randomly distributed fibres lead to diagonal components equal to 1/3. Moreover, the sum of the matrix diagonal is always 1. The orientation tensor of the extruded part has a strong preferential along the z-axis. The dominant circular fibre orientation in the injection moulded ring yields an almost planar random distribution as quantified by the orientation.

The porosity of fibre-reinforced polymers is a very important factor for the practical applications of this material since there is a direct correlation between porosity and mechanical properties, such as shear strength [49]. Therefore the porosity values of carbon fibre-reinforced composites for practical use in aircrafts and sports must be lower than 2.5–5 vol.%. The most common non-destructive method for measuring the porosity is ultrasonic testing. However, the ultrasonic attenuation coefficient depends not only on the porosity, but also on the shape and distribution of the pores and the presence of other heterogeneities. This can lead to significant errors in the localisation of the pores.

XCT is a very promising method for the non-destructive determination of the spatial distribution of porosity in composites. For this purpose it is important to use an appropriate segmentation method to find the correct surface between material and porosity. An example of what the microstructure of porous carbon fibre reinforced polymer (CFRP) samples for the aeronautic industry looks like can be found in Fig. 9, where an XCT-tomogram together with an optical picture of a

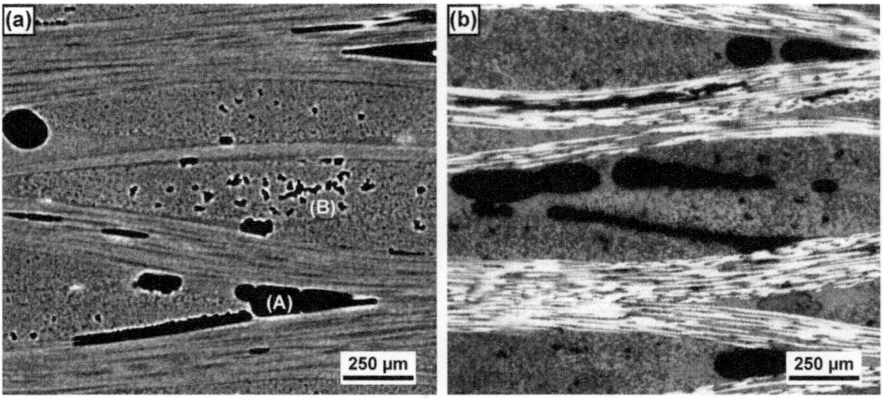

Fig. 9 Pores in a carbon fibre reinforced polymer sample with a porosity of ~4.6%: **a** XCT-tomogram with marked macro-pores (*A*) and micro-pores (*B*), voxel size = (2.75 µm)^3; **b** light optical micrograph

similar region of the sample are shown. The porosity in that sample amounts to 4.60 ± 0.04 vol.% determined by a segmentation method based on global thresholding. Pores bigger than 600 μm^3 can be registered in size and location. The contrast at the borders is somewhat enhanced. The average porosity values do not differ more than 0.3 percentage points (relatively 7%) from the porosity values obtained by the standard methods of ultrasonic testing and of acid digestion [50].

3.2 Light Metals

There are many publications on the subject of XCT of light metals. As an example, this section shows the possibilities of XCT applied to high strength AlZnMgCu-alloys. High strength wrought aluminium alloys are widely used in aerospace vehicles because of their high specific stiffness and strength, good workability and resistance to corrosion. The as–cast microstructure of the pre-material for rolling, extrusion and forging basically consists of α-Al dendrites and interdendritic eutectic regions due to segregations during solidification. Sc and Zr are added to aluminium alloys to form $Al_3(Sc,Zr)$ dispersoids. These dispersoids serve mainly to inhibit grain growth, resulting in improved strength and toughness and may activate grain nucleation if added in sufficient amounts [51–54].

The investigated chill cast Al–Zn–Mg–Cu–billets contain Sc and Zr, which form secondary $Al_3(Sc,Zr)$ dispersoids during homogenisation, but also primary $Al_3(Sc,Zr)$ in the melt. In addition to these particles, interdendritic segregations rich in highly absorbing elements like Zn and Cu appear.

Figure 10a, b show tomograms of CT–measurements with different measurement parameters, voxel sizes and sample geometry taken from an as-cast billet of Al-Zn-Cu-Mg. The interdendritic eutectic (containing Zn_2Mg and Cu-rich phases) and $Al_3(Sc,Zr)$ dispersoids are higher-absorbing than the α-Al–dendrites and thus appear brighter. In Fig. 10a a high-absorbing area at the outer edge of the slab with a width of approximately 0.45 mm can be identified in the tomogram. Zn and Cu are distributed more homogenously in that region than in the rest of the sample. Consequently, it can be deduced that fine grains with a relatively high solid solution of the alloying elements are solidified close to the surface.

Figure 10b presents the cell structure of the as-cast condition with higher resolution due to the smaller sample diameter. Bright spots as well as black pores are visible.

The low-absorbing pores have been segmented by a region growing-algorithm in a cuboid of $0.7 \times 1.2 \times 2.8$ mm^2 to produce Fig. 10c. Their measured volume fraction is about 0.3 vol.%. Due to the alignment of the principal axis of these pores the direction of solidification of the α-Al dendrites is traceable and averages at around 45° with respect to the surface of the billet.

Both the eutectic region and the $Al_3(Sc,Zr)$-particles similarly absorb more X-rays than the α-Al dendrites. Because of the limited spatial resolution and the appearance of physical artefacts, it was not possible to differentiate between these

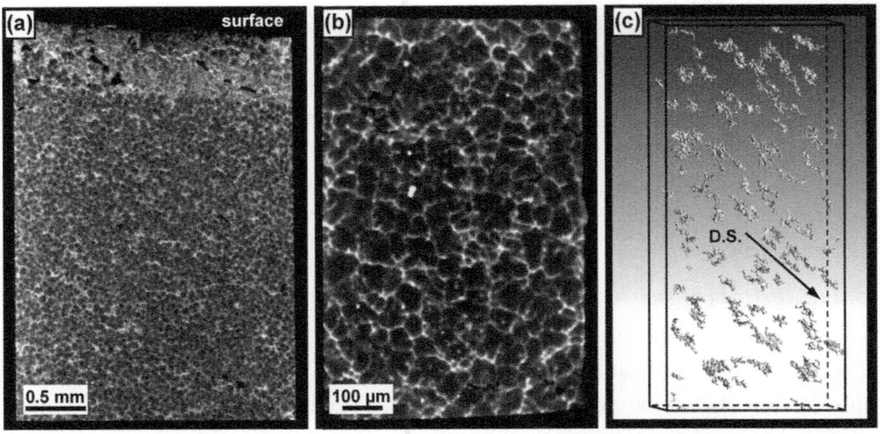

Fig. 10 Tomograms of a sample taken from an as-cast AlZnCuMg-billets alloyed with 0.11 wt% Zr, 0.25 wt% Sc: **a** slice of an $AlZn_8Cu_2Mg_2$–alloy with high absorbing regions as well as less absorbing spots (pores) between α-Al dendrites in a sample with 2.2 × 2.5 mm^2 cross section, voxel size: $(2.3\ \mu m)^3$. **b** slice of an $AlZn_6Cu_2Mg_2$-alloy showing the interdentritic cell structure in a sample with 0.7 × 1.2 mm^2 cross section, voxel size: $(1.5\ \mu m)^3$. **c** 3D view of micro-pores in a transparent volume of 0.7 × 1.2 × 2.8 mm^2 indicating the direction of solidification (D.S.), voxel size : $(1.5\ \mu m)^3$

two types of heterogeneities by means of the contrast level. Since the $Al_3(Sc,Zr)$–particles exhibit more or less equi-axial shapes, it is feasible to separate them from the irregular eutectic region by means of a shape factor. Thus, the shape factor F which represents the three-dimensional sphericity is introduced and calculated by Eq. 2 for each individual heterogeneity [55].

$$F = 6 \cdot \sqrt{\pi} \cdot \frac{volume}{\sqrt{surface^3}} \text{ where } F \in [0, 1], \text{ ideal sphere } F = 1. \tag{2}$$

A suitable threshold of $F = 0.4$ was applied on a volume of 1.09 × 0.73 × 0.14 mm^3. 190 almost spherical $Al_3(Sc,Zr)$—particles were distinguished from the eutectic region as illustrated in Fig. 11a. The volume fraction of the $Al_3(Sc,Zr)$—particles bigger than 5 μm in diameter amounts to 0.17 vol.%, their number density is > 1,500/mm^3 and they are centred within dendrites. Figure 11b shows three orthogonal tomogram slices of a typical α-Al dendrite (grey) where $Al_3(Sc,Zr)$—particles are located in the centre of the dendrite. Typical diameters of the star-shaped dendrites are about 200 μm whereby the arm width can be up to 30 μm. The position of the several μm-sized equi-axed $Al_3(Sc,Zr)$–particles indicates that they act as seed crystals for the α-Al dendrites and are segregated primarily from the melt. The metallurgist can deduce the amount of primary $Al_3(Sc,Zr)$ and how much Sc and Zr remains for precipitation hardening [56].

The morphology and distribution of the interdendritic eutectic replicates the dendritic solidification structure, where the orientation and width of the individual dendrite arms can be measured. Conclusions about the speed and direction of the

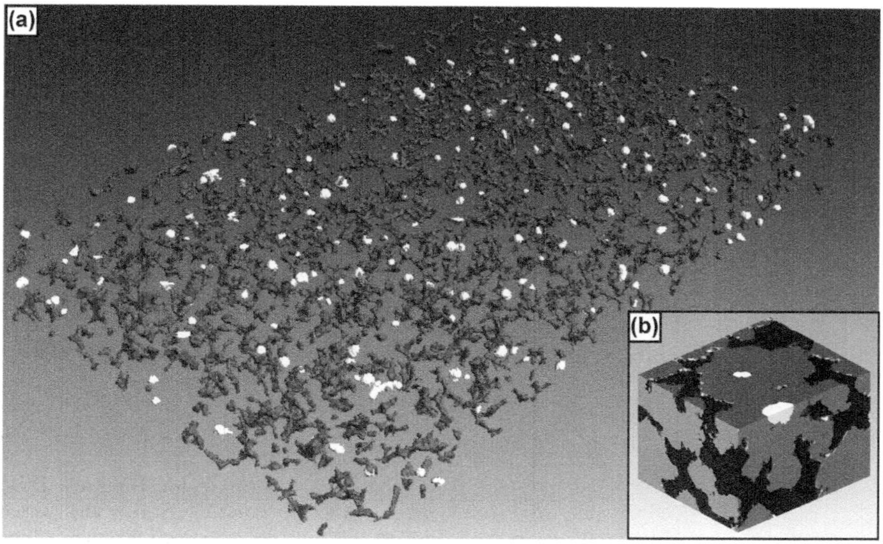

Fig. 11 Differentiation between $Al_3(Sc,Zr)$–particles (*white*) from the interdendritic eutectic region (*dark*): **a** volume of $1.09 \times 0.73 \times 0.14$ mm^3 in 3D. **b** typical arrangements of α-Al dendrites (*grey*), eutectic regions (*dark*) and $Al_3(Sc,Zr)$–particles (*white*) in a cuboid of $0.09 \times 0.11 \times 0.07$ mm^3 where $Al_3(Sc,Zr)$–particles are located centrally in the α-Al dendrites

solidification process can be drawn from these results at different positions in the slab. The morphology and the extension of the interdendritic segregations can be determined in 3D with high accuracy in different positions of a casting within a relatively large volume (in comparison to 2D metallographic techniques).

To a large extent there is a linear correlation between the detectability of heterogeneities and the voxel size. It is possible to identify the heterogeneities in aluminium alloys with a diameter of at least three voxels for reliable detection (i.e. a volume > 30 voxels).

3.3 Fe-Based Materials

3.3.1 Ductile Cast Iron

Ductile cast iron is characterised by high strength at high temperatures, good castability, durability, and damping behaviour and has numerous areas of application, e.g. in economic product design in the automotive industry. Apart from casting defects, the size and distribution of spherical graphite particles influences the mechanical properties considerably. The mean particle distance and the size distribution are important material characteristics and enable valuable conclusions about the solidification process.

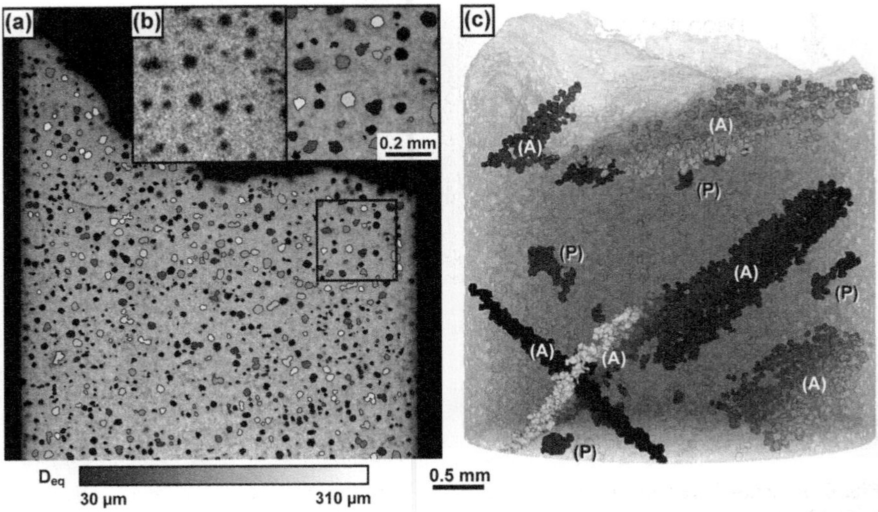

Fig. 12 Quantification of graphite particles in a fractured tensile sample of ductile iron (GJS 500): **a** slice of a tomogram with segmented graphite particles (*grey value coded* according to their size D_{eq}) with enlarged *inset* (**b**) comparing the segmented with the measured contrast, **c** spatial arrangement of graphite alignments (*A*) and some micro-pores > 50 μm (*P*)—iron matrix semi-transparent

As an example for the quantitative evaluation of spherical graphite particles, a fractured cylindrical tensile sample with a diameter of 4.5 mm was imaged. The voxel size in the CT-dataset is $(5.0\ \mu m)^3$ and graphite particles > 15–20 μm can be identified. The distribution of particles can be determined by a quantitative 3D analysis: 3D filters and special segmentation routines [57, 58] were applied in a volume of approximately 40 mm^3. In that volume more than 33,000 particles with a spherical equivalent diameter (D_{eq}) between 30 and 310 μm were analysed automatically. Fig. 12a shows a length section of a tomogram with the segmented particles, which are grey value coded according to their size. The enlargement in Fig. 12b shows the quality of the segmentation (right) in comparison to the corresponding original CT-measurement tomogram (left). A good correlation even for the smaller and noisier particles can be found so that the determined volume fraction of about 13 vol.% is quite reliable.

The VDG P441 [59] or EN ISO 945 standard [60] provides information about the method of graphite classification. The diameter of about 12 % of the segmented graphite particles exceeds 60 μm and in accordance with the specification, these particles fall into class 5 which degrades the quality significantly.

Alignments of graphite particles can be identified in 3D (Fig. 12c) by areas where the particle density is noticeably higher than in the rest of the sample. A segmentation of these particles shows dome-shaped structures along distinct planes, which might influence the toughness. In Fig. 12c six clusters of alignments marked with (A) and additionally occurring micro-pores > 50 μm (P) are shown in a semi-transparent 3D view. The effects of graphite alignments and micro-pores on crack propagation during tensile loading can be assessed by applying XCT-investigations.

3.3.2 Steel Slabs

The detection of heterogeneities in steels after different processing steps is of intrinsic importance to evaluate their classification as defects and their harmfulness in the process and for application. The spatial distribution of shrinkage pores, gas pores and non-metallic inclusions in steels can be represented by XCT over the entire slab cross section. Both the size of the heterogeneity and the size of the sample require an appropriate choice of voxel size achievable by the tomographic method.

In Fig. 13a, the alignment of shrinkage pores in the centre of a continuously cast steel slab and the typical shape of an individual segmented shrinkage pore are shown. The branched structure of the pore results in a shape factor F (see Eq. 2) near to zero. Complementary dendrite characteristics like primary and secondary arm distances can be obtained by the spatial arrangement of these heterogeneities in 3D. These characteristics can be used to reach conclusions about the direction and speed of solidification as well as about the 3D melt flow during continuous casting [61].

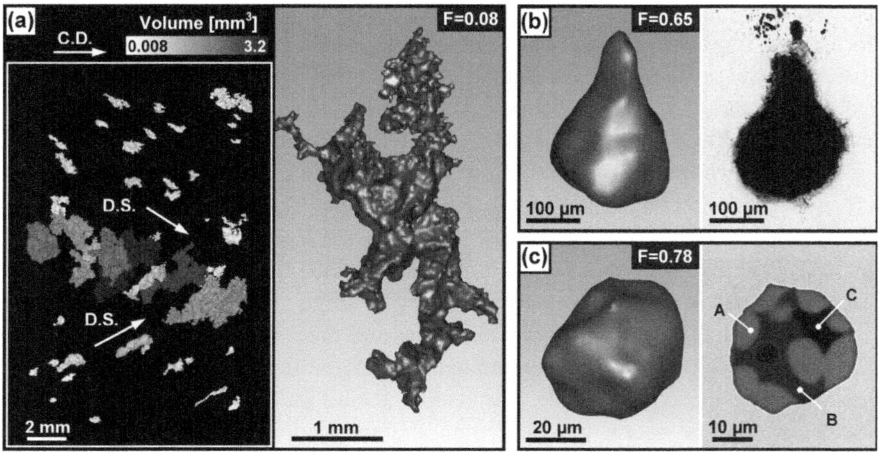

Fig. 13 3D illustrations of detectable heterogeneities in steel with shape factors F (sphericity) and their metallographic verification: **a** *Left*: alignment of shrinkage pores (*grey value* coded according to the volume of the pores) with casting direction (C.D.) and the directions of solidification (D.S.), *right*: enlarged shrinkage pore. **b** *Left*: gas pore, *right*: corresponding light-optical-micrograph. **c** *Left*: non-metallic multiphase inclusion, *right*: scanning-electron-microscope picture with energy-dispersive X-ray identification (EDX) of the phases: *A*: Ca-sulphides, *B*: Ca-aluminates, *C*: Spinel-types ($MgO.Al_2O_3$)

The 3D shape of a gas pore and the verification of the origin of the CT-contrast achieved by target preparation for metallography are shown in Fig. 13b. The mass density of oxide and sulphuric inclusions is about half the density of iron providing a sufficient difference in the X-ray absorption coefficient to quantify these particles, as well as shrinkage/gas pores bigger than the resolution limit of the XCT-system applied. A multiphase inclusion with a diameter of about 40 μm and the corresponding scanning-electron-microscope (SEM) picture is shown in Fig. 13c. Oxides of the inclusions can be identified by energy-dispersive X-ray analysis (EDX) in a SEM.

Non-metallic inclusions in steels define the degree of steel cleanliness, which is of increasing importance for high-quality steelmaking. The testing methods can be divided into two groups:

- Statistical methods such as light-optical-microscopy (or SEM) which cover the range of very small sized "micro inclusions" (typically < 10 μm, but also up to 100 μm). These methods are applied to 2D sections of small samples usually taken from the final product.
- Ultrasonic testing which usually covers the range of "macro inclusions" (typically > 100 μm). This method detects heterogeneities within the volume of wave penetration and can be applied to larger samples or even to the whole product (100% control).

The size range between 10 and 100 μm has recently gained increasing interest [62–64] as improved steelmaking methods have reduced the number and size of classical macro-inclusions. Medium-sized inclusions in the range of a few particles per mm^3 have a significant effect on aspects of product quality such as toughness, formability and fatigue strength. The conventional method for the detection of these inclusions is randomly located 2D metallographies, where the problem is the statistical confidence level because of the small area (volume) which can be evaluated within a reasonable time.

A sample was taken about 80 mm from the surface of an experimental slab. The cross section of the sample was 3.9×3.9 mm^2 and a preliminary inspection with an automated SEM-EDX cleanness analysis system showed oxide and sulphide inclusions with diameters up to 40 μm. By using XCT at a voxel size of (6.1 μm)^3, heterogeneities in the size range from 30 to 500 μm could be found (see Fig. 14a).

Because of the limited spatial resolution and the appearance of physical artefacts, it was not possible to differentiate between non-metallic inclusions and pores by means of the contrast level. Analysing metallography and the CT-data, the inclusions appear more spherical than the micro-pores. Thus, a shape factor F which describes the sphericity and calculated by Eq. 2 was used for this discrimination. The result of applying a suitable threshold of $F = 0.5$ is shown in Fig. 14b. The size distribution of the spherical inclusions with a volume fraction of 0.009 vol.% is illustrated in Fig. 14c.

Tomographic methods can be applied to detect non-metallic inclusions in steels [65]. Medium (> 10 μm) and macro-sized inclusions (> 100 μm) can be identified reliably with micro-focus XCT. Micro inclusions (< 10 μm) are detectable with

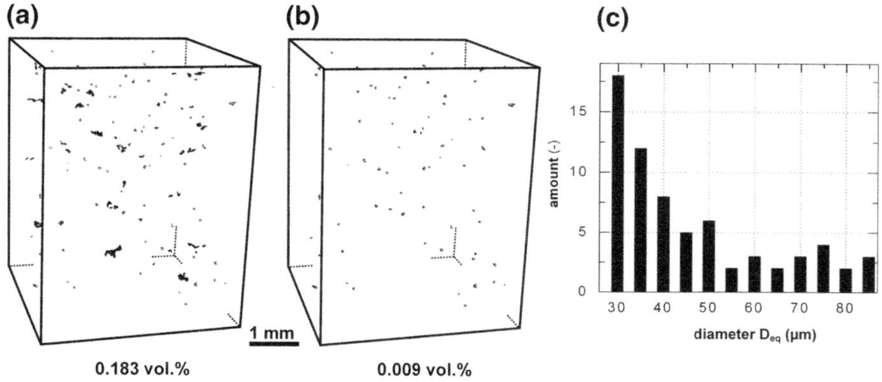

Fig. 14 Estimation of vol.% and the amount of non-metallic inclusions in a volume of ~80 mm^3 using the sphericity *F*: **a** 3D representation of pores and inclusions, **b** only inclusions F > 0.5, **c** size distribution of inclusions

sub-micro-focus XCT or synchrotron XCT. Due to the spherical shape of Ca-aluminates, detection of such inclusions with ultrasonic testing is difficult: using a shape factor in the CT-volume dataset is an appropriate method to differentiate between shrinkage pores and non-metallic inclusions. Thus, size distributions and inclusion contents over volumes of several 100 mm^3 can be determined. Samples of more than 80 mm^3 (4 mm Ø) with inclusions > 30 μm can be identified in one CT-scan (with a measurement time of about 1 h), which would be equivalent to approximately two hundred cross sections for 2D analysis by metallographic methods.

The combination of different XCT techniques covers a large and, for steel manufacturers, relevant range of possible sizes of heterogeneities in continuously cast steel slabs for an appropriate evaluation of cleanliness.

3.3.3 Sintered Metals

Sintering is a processing technique for making dense materials from powder by heating the material until its particles adhere to each other and the inner surfaces are reduced by diffusion. This technology is often applied to produce ceramic objects and in the field of powder metallurgy. Selective laser sintering (SLS) is developed to fabricate complex geometries directly from CAD data. In this additive manufacturing process the laser selectively melts powdered material by scanning across the surface of a powder bed [66].

The variation of the processing parameters has a significant influence on the porosity of the components. Therefore XCT can be applied in order to analyse the quality, to improve the density of a part and to adjust the manufacturing time. Cylindrical samples were produced with a diameter of approximately 5 mm and with different fabrication parameters by SLS layer by layer from steel powder

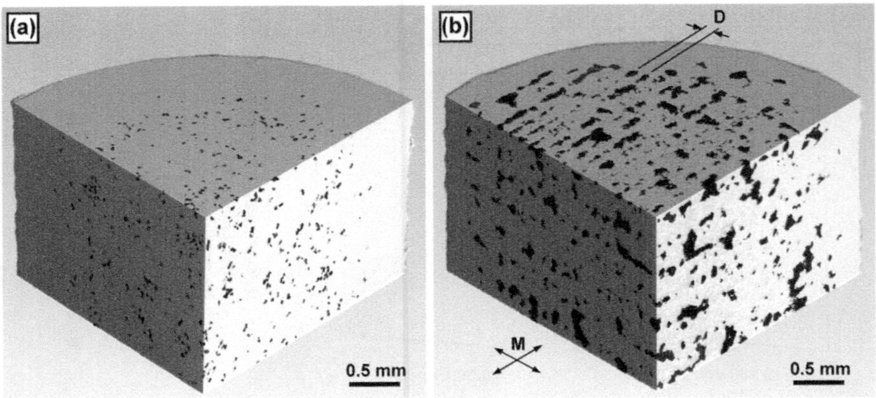

Fig. 15 3D presentation of pores (*black*) in selective laser sintered steel powder, manufactured with different laser scanning speeds: **a** $v = 200$ mm/s, porosity = 0.9 vol.%, **b** $v = 600$ mm/s, 7.8 vol.%. The distance (*D*) of laser-lines and alternate movements in layers (*M*) of the laser are marked

(X2CrNiMo17-12-2 with a powder grain size of 35 μm). XCT-measurements with a voxel size of $(7.5\ \mu m)^3$ were performed and the porosity values were determined by special segmentation routines for pores > 45 μm. In Fig. 15a three orthogonal XCT-slices with the segmented pores (black) are shown. This sample was manufactured with a laser scanning speed of 200 mm/s and the estimated porosity is only about 0.9 vol.%. The triple laser scanning speed leads to higher porosity values (7.8 vol.%) which can be seen in Fig. 15b. The pores are orientated in a way that allows reconstruction of the laser paths.

The densities were calculated using the CT-porosity values of more than 20 samples. Gravimetric density measurements yield a standard deviation of ±0.027 g/cm^3, whereas the accuracy of CT-evaluation amounts ±0.04 g/cm^3 in absolute terms or ±0.5% of the porosity [67].

3.4 Metallic Foams and Metal Matrix Composites (MMC)

Cellular metals (metallic foams) are a class of materials with low densities and an outstanding combination of mechanical, electrical, thermal and acoustic properties. They offer great potential in the areas of light-weight structures, energy absorption and thermal management. Foamed metals result from nucleation and expansion of gas bubbles in a liquid or semi-liquid metal (e.g. Al or Mg). They usually have a non-uniform pore structure with variable pore size and preferred orientation of the pores. The most important parameters to characterise a cellular structure are the relative density, the morphology of the cell, the topology and the properties of the cell wall material. Although resolving the structure of cellular foams has been the subject of scientific research for some time, no simple standard

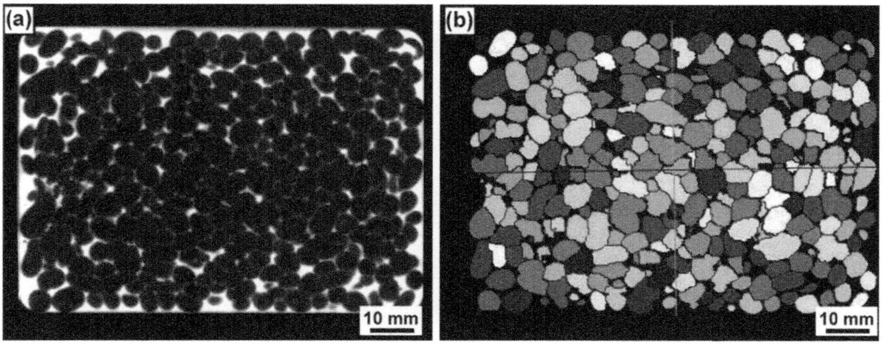

Fig. 16 Quantification of pores in foamed aluminium: **a** tomogram, voxel size: (232 μm)^3; **b** final Watershed image with removed border objects: the various coloured pores can be distinguished. The borders of the objects are the "watersheds" and they correspond to the aluminium walls between the pores

technique or procedure to determine some of the most common structural parameters has been identified to date [68–70].

The application of XCT in combination with adequate image processing seems to be an accurate and fast method to characterise foams. Figure 16a shows a tomogram of a closed-cell aluminium foam sample with a cross section of 65×92 mm^2. The raw datasets obtained from the CT-scans were processed using the commercial image processing software package MAVI-Modular Algorithms for Volume Images [71]. MAVI contains many algorithms for use on 3D images, such as filters, segmentation algorithms and morphological transforms, especially for the analysis of images of microstructures as found in materials like concrete or foams. This special image processing software was used to isolate pores within the foam and to calculate their volumes, mean diameters, surface area and sphericity.

The four main steps for the characterisation of the pores are filtering, binarisation, segmentation and the calculation of the individual object features. After applying Gaussian and Laplace filters a binarisation using the Otsu's method [41] was carried out. Then a Watershed algorithm was applied using the original CT-grey value image and the binary image to find and segment the pores. The final image after removal of the border objects for the aluminium foam is shown in Fig. 16b. At the end of this analysis procedure the object features for each individual pore can be generated. This information includes, e.g. pore diameters, volumes, surface areas, different shape factors (sphericity) and location within the specimen.

Metal matrix composites (MMC), like all composites, consist of at least two distinct phases, suitably distributed to provide properties not obtainable with either of the individual phases. Generally there are two phases, e.g. a fibrous or particulate ceramic phase, embedded in a metallic matrix. These composites combine, for example, high tensile strength or thermal and electrical conductivity with low specific weight but the properties of fibre reinforced materials are strongly

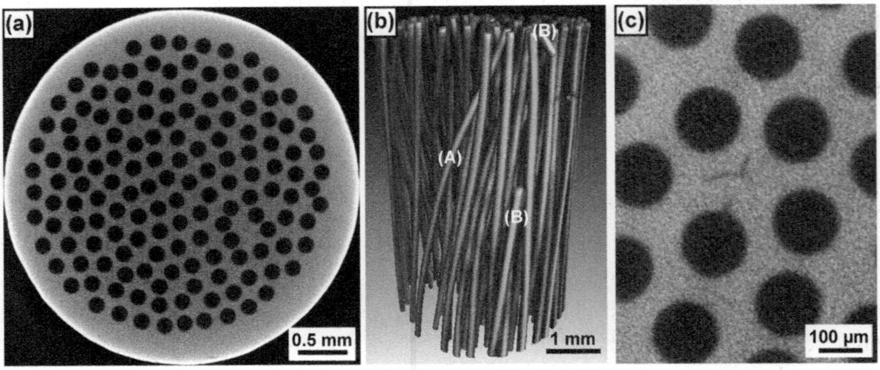

Fig. 17 Tomographic presentation of a CuCrZr/SiC/20 m sample: **a** tomogram with dark monofilaments, voxel size: (2.1 µm)^3. **b** 3D view of the segmented SiC-fibres with a bended fibre (*A*) and fibre fragmentations (*B*). **c** Enlarged section of the tomogram in (**a**) showing micro-cracks (*dark*) in the Cu-matrix between the reinforcement

influenced by fibre orientation and the interface between the fibre and the matrix as well as the distribution of cracks and pores [72, 73].

High performance MMCs based on Cu and its alloys reinforced by ceramic monofilaments such as SiC and Al_2O_3 are investigated for heat sink applications [74]. The thermal and mechanical properties of such MMCs largely depend on the reinforcement alignment and on the bonding strength between the fibres and the matrix. XCT can be applied in order to investigate the distribution, misorientations and fragmentations of the fibres, fibre/matrix-debonding or splicing of the fibres as well as micro-cracks and pores in the metallic matrix. Figure 17 shows some of these defects in a monofilament reinforced Cu-matrix (CuCrZr/SiC/20 m) with a sample diameter of 3.5 mm. The calculated volume fraction of the segmented SiC-fibres amounts ~19 vol.%.

4 Conclusions

This chapter describes investigations on the detection and characterisation of heterogeneities in materials by means of X-ray computed tomography methods, namely cone beam XCT with micro-focus and sub-micro-focus X-ray sources. The main conclusions are:

1. In order to guarantee a representative 3D description of heterogeneities, sampling is important. Both the size of the heterogeneity and the diameter of the sample require an appropriate choice of tomographic method (spatial resolution of the magnitude > diameter/1,000). There is a strict hierarchical structure between the analysis volume and the obtained spatial resolution.
2. The distribution of heterogeneities > 3 voxels in diameter can be determined reliably by a quantitative 3D analysis. Accumulations, frequency and positions

of such critical heterogeneities can be identified in statistically significant sample volumes in comparison to metallographic analysis.
3. Applying 3D shape factors to the detected heterogeneities is an appropriate method to discriminate between different types of heterogeneities. Thus, size distributions can be determined. Destructive methods are needed to identify the phases.
4. XCT methods are applicable to nearly every material system. For an appropriate application of tomographic methods knowledge of the limitations in the detectability of heterogeneities is of crucial importance. Therefore the relationship between analysis volume and the size of heterogeneity has to be considered. Regarding grey value contrasts $> 5\%$ in the CT-dataset, there is to a large extent a linear connection between the size of a heterogeneity and the spatial resolution (voxel size). In aluminium alloys the heterogeneities must be illustrated to a diameter of at least 3–5 voxels for reliable detection independent of the applied spatial resolution. Due to the occurrence of artefacts in high-absorbing materials, e.g. Fe-based alloys, heterogeneity diameters of at least 4–7 voxels are necessary. Furthermore, the position of the heterogeneities within the sample is important: the detectability of heterogeneities in the centre is worse and their morphology is geometrically distorted in comparison to heterogeneities which are located nearer to or at the edge of the sample.
5. Apart from the detection of pores and cracks, it is possible to quantify segregations and inclusions in cast metals in 3D. Their morphology and their distribution replicate the dendritic solidification structure, where the orientation and width of the individual dendrite arms can be measured. These characteristics can be used to draw conclusions about the direction and speed of solidification as well as about the 3D turbulent current during casting.

Acknowledgments The authors gratefully acknowledge the contributions of: Professor H.P. Degischer and his colleagues from the Vienna University of Technology, Institute of Materials Science and Technology for their support and useful input; M. Reiter, B. Plank, D. Salaberger and P. Orgill from the Upper Austria University of Applied Sciences; metallographic investigations of non-metallic inclusions by voestalpine Stahl AG; the provision of samples by the companies AMAG AG, FACC AG, Georg Fischer AG, Gruber & Kaja GmbH, Honsel AG, MPI-IPP Garching and TCKT GmbH. The project was supported by the COMET program of the FFG and Austrian research funds (FWF).

References

1. Hsieh, J.: Computed Tomography, Principles, Design, Artifacts and Recent Advances. SPIE The International Society for Optical Engineering, Bellingham (2003)
2. Buzug, T.M.: Computed Tomography: From Photon Statistics to Modern Cone-Beam CT. Springer, Berlin/Heidelberg (2008)
3. Oster, R.: Computed tomography as a non-destructive test method for fibre main rotor blades in development, series and maintenance. In: Proceedings of International Symposium on

Computerized Tomography for Industrial Applications and Image Processing in Radiology. DGZfP-BB 67-CD, Berlin (1999)
4. Illerhaus, B., Goebbels, J., Kettschau, A., Reimers, P.: Non destructive waste form and package characterization by computed tomography. Mat. Res. Soc. Symp. Proc. **127**, 507–512 (1989)
5. Habermehl, A., Ridder, H.W.: Ein neues Verfahren zum Nachweis der Rotfäule. In: International Conference on Problems of Root and Butt Rot in Conifers, pp. 340–347, Kassel (1978)
6. The Chinese Society of NDT (eds.): Proceedings of 17th World Congress on Non-destructive Testing. ChSNDT, Shanghai (2008)
7. Moseley, H.G.J.: The high-frequency spectra of the elements. Phil. Mag. **26**, 1024–1034 (1913)
8. Ebel, H.: Fundamental parameter programs: algorithms for the description of K, L and M spectra of X-ray tubes. Adv. X-ray Anal. **49**, 267–273 (2006)
9. Moore, P.O. (ed.): Nondestructive Testing Handbook, Radiographic Testing, vol. 4. American Society for Nondestructive Testing, Columbus (2002)
10. Eisberg, R., Resnick, R.: Quantum Physics of Atoms, Molecules, Solids, Nuclei, and Particles. John Wiley & Sons, New York (1985)
11. Lambert, J.H.: Photometria, sive de mensura et gradibus luminis colorum et umbrae. Sumptibus Vidae Eberhardi Klett, Leipzig (1760)
12. Beer, A.: Bestimmung der Absorption des rothen Lichts in farbigen Flüssigkeiten; von Beer. Ann. Phys. **86**, 78–87 (1852)
13. Berger, M.J., Hubbell, J.H., Seltzer, S.M., Chang, J., Coursey, J.S., Sukumar, R., Zucker, D.S.: National Institute of Standards and Technology Standard Reference Database 8 (XGAM). http://physics.nist.gov/PhysRefData/Xcom/Text/XCOM.html (2009). Accessed 8 June 2009
14. Baruchel, J., Buffiere, J.Y., Maire, E., Peix, G. (eds.): X-ray Tomography in Material Science. Hermes Science Publications, Paris (2000)
15. VDA Unterausschuss ZfP und DGZfP Fachausschuss Durchstrahlungsprüfung (eds.): VDA-Prüfblatt 236-101/DGZfP Merkblatt D6: Anforderungen und Rahmenbedingungen für den Einsatz der Röntgencomputertomographie in der Automobilindustrie. DGZfP, Berlin (2008)
16. Partridge, M., Hesse, B.M., Müller, L.: A performance comparison of direct- and indirect-detection flat-panel imagers. Nucl. Instrum. Methods Phys. Res. A **484**, 351–363 (2002)
17. Hoheisel, M.: Amorphous silicon X-ray detectors. J. Non Cryst. Solids **227–230**, 1300–1305 (1998)
18. Lee, H.R., Ellingson, W.: Characterization of a flat panel amorphous Si detector for CT. J. Xray Sci. Technol. **9**, 43–53 (2001)
19. Gondrom, S., Schröpfer, A.: Digital computed laminography and tomosynthesis-functional principles and industrial applications. In: Proceedings of International Symposium on Computerized Tomography for Industrial Applications and Image Processing in Radiology. DGZfP-BB 67-CD, Berlin (1999)
20. Simon, M., Sauerwein, C., Tiseanu, I.: Extended 3D CT method for the inspection of large components. In: Proceedings of 16th World Conference on Nondestructive Testing, ICNDT, Montreal (2004)
21. Feldkamp, L.A., Davis, L.C., Kress, J.W.: Practical cone beam algorithm. J. Opt. Soc. Am. A **6**, 612–619 (1984)
22. Reiter, M., Heinzl, C., Salaberger, D., Weiss, D., Kastner, J.: Study on parameter variation of an industrial computed tomography simulation tool concerning dimensional measurement deviations. In: Proceedings of 10th European Conference on Non-destructive Testing. RSNTTD, Moscow (2010)
23. Smith, B.D.: Image reconstruction from cone-beam projections: necessary and sufficient conditions and reconstruction methods. IEEE Trans. Med. Image **4**(1), 14–25 (1985)
24. Salvo, L., Cloetens, P., Maire, E.: X-ray micro-tomography an attractive characterisation technique in materials science. Nucl. Instrum. Methods Phys. Res. Sect. B **200**, 273–286 (2003)

25. SkyScan: Micro-CT in SEM. http://www.skyscan.be/products/SEM_microCT.htm. Accessed 29 June 2010
26. Perkin Elmer Inc.: Reference Manual. http://www.perkinelmer.com (2009). Accessed 26 March 2009
27. Simon, M., Sauerwein, C., Tiseanu, I., Burdairon, S.: Flexible 3D-Computertomographie mit RayScan 200. In: Proceedings DGZfP Annual Conference, DGZfP, Berlin (2001)
28. RayScan Technologies GmbH: RayScan 250. http://www.rayscan.eu/PDF-Internet/PRO-G75022-Be-i.pdf (2009). Accessed 25 March 2009
29. Paul, T., Zhenhui, H.E.: Advanced NDT with high resolution computed tomography. In: Proceedings of 17th World Conference on Non-destructive Testing, ChSNDT, Shanghai (2008)
30. Hamamatsu Photonics K.K. Flat panel sensor C7942SK-05. http://sales.hamamatsu.com/ (2009). Accessed 28 March 2009
31. Krimmel, S., Stephan, J., Baumann, J.: 3D computed tomography using a microfocus X-ray source: analysis of artefact formation in reconstructed images using simulated as well as experimental projection data. Nucl. Instrum. Methods Phys. Res. Sect. A **542**, 399–407 (2005)
32. Herman, G.T.: Correction for beam hardening in computed tomography. Phys. Med. Biol. **24**(1), 81–106 (1979)
33. Hopkins, F., Du, Y., Lasiuk, B., Abraham, A., Basu, S.: Analytical corrections for beam-hardening and object scatter in volumetric computed tomography systems. In: Proceedings of 16th World Conference on Nondestructive Testing, ICNDT, Montreal (2004)
34. Hammersberg, P., Mangard, M.: Correction for beam hardening artefacts in computerised tomography. J. Xray Sci. Technol. **8**, 75–93 (1998)
35. Van de Casteele, E.: Model-based approach for beam hardening correction and resolution measurements in microtomography. PhD Thesis, University of Antwerp, Antwerp (2004)
36. Kasperl, S.: Qualitätsverbesserungen durch referenzfreie Artefaktreduzierung und Oberflächennormierung in der industriellen 3D-Computertomographie. PhD Thesis, Technische Fakultät der Universität Erlangen, Nürnberg (2005)
37. Joseph, P.M., Spital, R.D.: The effects of scatter in X-ray computed tomography. Med. Phys. **9**(4), 464–472 (1982)
38. Wiegert, J.: Scattered radiation in cone-beam computed tomography: analysis, quantification and compensation. PhD Thesis, Rheinisch-Westfälischen Technischen Hochschule Aachen, Aachen (2007)
39. Banhart, J. (ed.): Advanced Tomographic Methods in Materials Research and Engineering. Research Oxford University Press, Oxford (2008)
40. Prewitt, J.M.S., Mendelsohn, M.L.: The analysis of cell images. Ann. N.Y. Acad. Sci. **128**(3), 1035–1053 (1966)
41. Otsu, N.: A threshold selection method from gray level histograms. IEEE Trans. Sys. Man. Cyber. **9**(1), 62–66 (1979)
42. Russ, J.C.: The Image Processing Handbook. CRC Press LLC, Boca Raton (2002)
43. Rosenfeld, A., Pfaltz, J.L.: Sequential operations in digital picture processing. J. ACM **13**(4), 471–494 (1966)
44. Heinzl, C., Kastner, J., Gröller, E.: Surface extraction from multi-material components for metrology using dual energy CT. IEEE Trans. Visual Comput. Graphics **13**(3), 1520–1528 (2007)
45. Kastner, J. (ed.): Proc. Industrielle Computertomografietagung, Wels. Shaker Verlag, Aachen (2008)
46. Kastner, J., Salaberger, D., Zitzenbacher, G., Stadlbauer, W., Freytag, R.: Determination of diameter, length and three-dimensional distribution of fibres in short glass-fibre reinforced injection moulded parts by X-ray-computed tomography. In: Proceedings of 24th Annual Meeting Polymer Processing Society, Salerno (2008)
47. Tan, J.C., Elliot, J.A., Clyne, T.W.: Analysis of tomography images of bonded fibre networks to measure distribution of fibre segment length and fibre orientation. Adv. Eng. Mater. **8**(6), 495–500 (2006)

48. Pfeifer, F., Kastner, J., Freytag, R.: Method for three-dimensional evaluation and visualization of the distribution of fibres in glass-fibre reinforced injection moulded parts by μ-X-ray computed tomography. In: Proceedings of 17th World Congress on Non-destructive Testing. ChSNDT, Shanghai (2008)
49. Birt, E.A., Smith, R.A.: A review of NDE methods for porosity measurement in fibre-reinforced polymer composites. Insight **46**(11), 681–687 (2004)
50. Plank, B., Kastner, J., Sekelja, J., Salaberger, D.: Determination of porosity in carbon fibre reinforced polymers by X-ray computed tomography. In: Proceedings of 4th FFH Conference, pp. 321–328. Rötzer, Eisenstadt (2010)
51. Mondolfo, L.F.: Aluminium Alloys: Structure and Properties. Butterworths, London (1976)
52. Davis, J.R. (ed.): Aluminum and Aluminum Alloys. ASM International, Ohio (1993)
53. Radmilovic, V., Tolley, A., Lee, Z., Dahmen, U.: Core-shell structures and precipitation kinetics of Al_3 (Sc, Zr) Li_2 intermetallic phase in Al-rich alloy. Metalurgija-J. Metallurgy **12**, 309–314 (2006)
54. Senkov, O.N., Shaghiev, M.R., Senkova, S.V., Miracle, D.B.: Precipitation of Al3(Sc, Zr) particles in an Al-Zn-Mg-Cu-Sc-Zr alloy during conventional solution heat treatment and its effects on tensile properties. Acta. Mater. **56**(15), 3723–3738 (2008)
55. Mücklich, F., Ohser, J.: Statistical Analysis of Microstructures in Materials Science. Wiley and Sons, Weinheim (2000)
56. Harrer, B., Degischer, H.P., Kastner, J.: Microfocus computed X-ray tomography of segregations in high strength aluminium alloys. In: Proceedings of 10th European Conference on Non-destructive Testing. RSNTTD, Moscow (2010)
57. Reinhart, C., Poliwoda, C., Guenther, T., Roemer, W., Gosch, C.: Modern voxel based data and geometry analysis software tools for industrial CT. In: Proceedings of 16th World Conference on Nondestructive Testing, ICNDT, Montreal (2004)
58. Volume Graphics GmbH. Reference Manual VGStudio Max Release 2.0. http://www.volumegraphics.com (2008). Accessed 28 March 2009
59. Association of German Foundryman (eds.): Merkblatt P441: Richtreihe zur Kennzeichnung der Graphitausbildung. VDG, Düsseldorf (1962)
60. ISO Standard 945-1:2008: Microstructure of cast irons—Part 1: Graphite classification by visual analysis. Geneva (2008)
61. Harrer, B.: Detectability of heterogeneities in Fe-based and aluminum alloys by X-ray computed tomography. PhD Thesis, Vienna University of Technology, Vienna (2009)
62. Meilland, R., Jowitt, R., Ribera, M., Didriksson, R., Schuller, M.: Research Project 218-PR/039, Improvement of sampling and analysis procedures for clean and high purity steel. EC, Luxembourg (2002)
63. Hansen, T., Runnsjö, G., Törresvoll, K., Jönsson, P.G.: On the assessment of macro inclusions in stainless steels using ultrasonic technique. In: Proceedings of 16th World Conference on Nondestructive Testing, ICNDT, Montreal (2004)
64. Boue-Bigne, F.: Laser-induced breakdown spectroscopy applications in the steel industry: rapid analysis of segregation and decarburization. At. Spectrosc. **63**(10), 1122–1129 (2008)
65. Harrer, B., Kastner, J., Winkler, W., Degischer, H.P.: Opportunities in the detection of inhomogeneities in steel by computed tomography. In: Proceedings of 17th World Congress on Non-destructive Testing. ChSNDT, Shanghai (2008)
66. Santos, E.C., Shiomi, M., Osakada, K., Laoui, T.: Rapid manufacturing of metal components by laser forming. Int. J. Mach. Tools Manuf. **46**, 1459–1468 (2006)
67. Plank, B., Kastner, J., Schneider, R., Busch, R.: Characterisation of porosities in sinter-metals by X-ray computed tomography-selective laser sintering. In: Proceedings DGZfP Annual Conference, pp. 810–816. DGZfP, Münster (2009)
68. Degischer, H.P., Kriszt, B. (eds.): Handbook of Cellular Metals. Wiley-VCH, Weinheim (2002)
69. Clyne, T.W., Simancik, F.: Metal Matrix Composites and Metallic Foams. Wiley-VCH, Weinheim (2000)

70. Scheffler, M., Colombo, P.: Cellular Ceramics: Structure, Manufacturing, Properties and Applications. Wiley-VCH, Weinheim (2005)
71. Fraunhofer-Institut für Techno- und Wirtschaftsmathematik. MAVI-Modular Algorithms for Volume Images. http://www.itwm.fhg.de/bv/projects/MAVI/ (2009). Accessed 12 June 2010
72. Chawla, N., Chawla, K.K.: Metal Matrix Composites. Springer, New York (2006)
73. Kainer, K.U. (ed.): Metal Matrix Composites. Custom-Made Materials for Automotive and Aerospace Engineering. Wiley-VCH, Weinheim (2006)
74. Brendel, A., Popescu, C., Köck, T., Bolt, H.: Promising composite heat sink material for divertor of future fusion reactors. J. Nucl. Mater. **367–370**, 1476–1480 (2007)

Submicron Tomography Using High Energy Synchrotron Radiation

András Borbély, Peter Cloetens, Eric Maire and Guillermo Requena

Abstract Development of synchrotron tomography at the end of the 20th century and the following improvements in radiation source, X-ray detectors as well as X-ray optics boosted the application of the tomographic technique in materials science. It became possible for the first time to reveal the three dimensional structure of heterogeneous materials with sub-micrometer spatial resolution, the length scale where the basic mechanisms of plastic deformation and damage are taking place and determine the macroscopic behavior of engineering components. The present chapter introduces the basic features of the tomographic method developed at synchrotron sources, related mainly to the high flux and lateral coherence of the beam. These allow performing high resolution tomographic scans within a reasonable time, but also to use the optical phase of the transmitted X-ray beam to increase the sensitivity by revealing the spatial distribution of electron density. After the introduction of the technique several application examples in the field of materials science are presented.

A. Borbély (✉)
MPM Department, SMS Center, École des Mines de Saint-Étienne,
158 cours Fauriel, 42023 Saint-Étienne cedex 2, France
e-mail: borbely@emse.fr

P. Cloetens
European Synchrotron Radiation Facility (ESRF), 38043 Grenoble, France
e-mail: cloetens@esrf.eu

E. Maire
Matériaux Ingénierie Science MATEIS, INSA de Lyon, 69621 Villeurbanne, France
e-mail: eric.maire@insa-lyon.fr

G. Requena
Institute of Materials Science and Technology, Vienna University of Technology,
Karlsplatz 13/308, Vienna, Austria
e-mail: grequena@mail.zserv.tuwien.ac.at

F. A. Lasagni and A. F. Lasagni (eds.), *Fabrication and Characterization in the Micro-Nano Range*, Advanced Structured Materials, 10,
DOI: 10.1007/978-3-642-17782-8_7,

1 Introduction

Synchrotron tomography became interesting for materials science studies when the spatial resolution of the technique reached the micrometer range [1, 2]. This allowed reconstructing three dimensional (3D) representative volumes of heterogeneous materials with sufficient microstructural details, necessary to predict the macroscopic properties. For example the permeability of porous Fontainebleau sandstone estimated on the basis of a tomographic reconstruction showed good agreement with experimental data [3]. Borbély et al. [4] have also shown that the effective elastic and plastic properties of a metal matrix composite, consisting of 6061 aluminium matrix alloy reinforced with 20 vol.% alumina particles, can be well predicted with the so called "*mean window technique*". The method is based on the finite element simulation of real material sub-volumes (windows) having a local volume fraction of particles equal to the mean volume fraction. As a requisite the edge length of such volumes should be larger than the two-point correlation length of the underlying structure.

High-resolution X-ray tomography [5, 6] with synchrotron radiation is excellently suited for observation of changes in the local microstructure. The easiest cases are related to damage identification in the form of pores or cracks, when the difference between X-ray attenuation coefficients of the matrix and the air is large. Obviously the high photon flux available at synchrotron source facilitates in situ tomographic investigations, but when microstructural changes are faster than the time necessary for the scan, measurements on interrupted test specimens can be also very helpful to understand the kinematics of studied processes. Results of interrupted test specimens revealed the fracture process of metal matrix composites [7], the early stages of fatigue crack nucleation in an austempered ductile cast iron [8]. The latter authors have shown that stable short cracks nucleated in the stress concentration zone of a pore and arrested due to microstructure barriers. Ductile damage of metals has also been studied using interrupted tensile tests in model metal matrix composites [9] and even in absorbent (steel) samples [10].

Recently the acquisition speed has increased allowing the in situ tests to be performed while the load is continuously applied to the sample. In situ tomography in continuous loading has been successfully applied to study creep damage in leaded brass [11] and also during continuous tensile tests [12]. In each of these cases, synchrotron tomography reveals the chronology of damage (distinct observation of the initiation, growth and coalescence of voids or micro-cracks) and permits to compare experimental results to classical modeling of these different phases.

These and other numerous observations demonstrate the excellent potential for future applications of high-resolution tomography to materials science research. The following paragraphs describe the physical basis of X-ray synchrotron tomography and some technical aspects of the setups as implemented at the European Synchrotron Radiation Facility (ESRF).

2 High Resolution Synchrotron Tomography

The usual source of contrast in X-ray tomography is the change in the attenuation properties of the material. Absorption tomography retrieves the distribution of the linear attenuation coefficient $\mu(x, y, z)$. This quantity strongly depends on material properties through the atomic number Z of the present elements. The rapid increase of the attenuation for heavy elements must be compensated by increasing the X-ray energy according to the characteristic cubic dependence:

$$\mu \propto (Z/E)^3. \tag{1}$$

Therefore, to investigate structural materials with relevant size on which in situ experiments can be performed, rather high X-ray energies are required: typically 10–20 keV for aluminium based samples, 20–40 keV for iron, copper or titanium based samples. Performing tomography with truly submicron resolution at higher energies remains far from straightforward.

In many relevant cases, the attenuation contrast is too weak to be exploited. This is particularly true for light materials (e.g., polymer based), at high energies and at high spatial resolution. Much improved sensitivity can be achieved in general by exploiting the optical phase of the transmitted beam instead of its amplitude. Phase contrast micro-tomography techniques have been developed at synchrotron facilities exploiting the coherence properties of the delivered beams. In this case the contrast is due to changes in the refractive index of the material, which is essentially determined by the electron density. Quantitative phase tomography therefore retrieves the electron density distribution in the sample. Figure 1b shows the relative strength of phase contrast versus attenuation contrast as a function of the X-ray energy. A gain of several orders of magnitude can be achieved, especially at higher energies. Several methods are employed to obtain phase contrast, such as propagation based, analyzer based and grating based phase contrast. In a microscope configuration, Zernike type phase contrast can be achieved. Propagation based and Zernike type phase-contrast are the most adapted to achieve sub-micron spatial resolution.

The most trivial approach of phase contrast tomography is propagation based as it occurs naturally as soon as there is a finite distance between the X-ray source, the sample and the detector. After propagation an interference pattern builds up that can be ideally resolved with synchrotron beams due the high spatial coherence or equivalently the exceptionally good geometric resolution. Propagation based phase contrast can be used as a simple edge-enhancement technique improving the detection of cracks, metallurgic phases with similar attenuation, etc. Quantitative phase tomography requires most often the acquisition of more data as compared to absorption tomography. In holotomography [13], a few (typically three or four) tomographic scans are acquired corresponding to different distances between the sample and the detector, or between the X-ray source and the detector. A numerically implemented holographic phase retrieval step is combined with the usual tomography procedure. In this case two independent 3D data sets of the

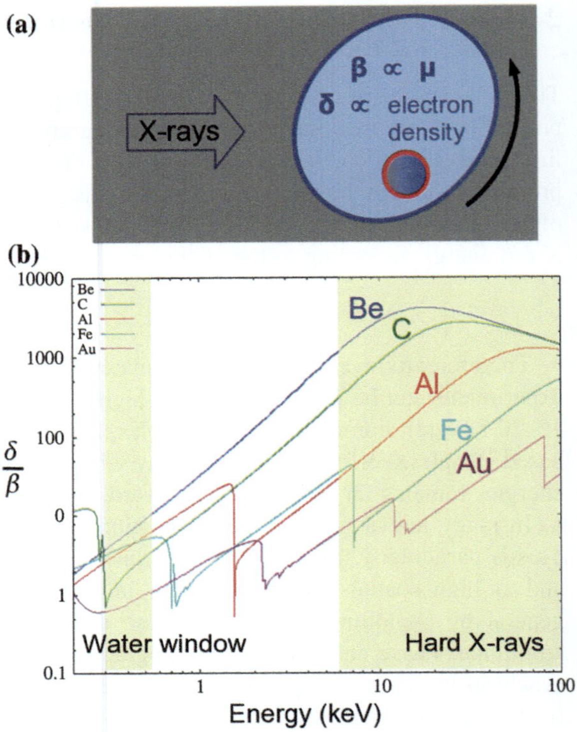

Fig. 1 **a** X-ray transmission imaging to map the distribution of the absorption index β (proportional to the linear attenuation coefficient μ) and the refractive index decrement δ, at the origin of phase contrast; **b** Relative strength of phase contrast versus attenuation contrast as a function of the X-ray energy

sample are produced: the distribution of the attenuation coefficient and the electron density distribution (Fig. 1a).

Other material properties and contrast mechanisms can also be exploited in X-ray imaging. X-ray fluorescence can be excited and used to obtain a chemical image with a signature of specific elements. X-ray fluorescence microscopy is typically performed with a focused beam in scanning geometry. Fluorescence tomography [14] can be achieved but it suffers from long acquisition times and from strong limitations on heavy materials due to self-absorption effects (the low energy fluorescence X-rays are re-absorbed by the specimen and can not leave it). Combination of tomography with X-ray diffraction is relevant to obtain information on the crystalline nature of the sample. It is used on poly-crystalline multiple phase materials in a (slow) scanning geometry (diffraction tomography [15]) or on single phase materials in full-field geometry (diffraction contrast tomography [16]). In diffraction tomography it is assumed that the grain size is much smaller than a resolution element, whereas for diffraction contrast tomography the grain size should be significantly larger than the resolution element.

Concerning spatial resolution, different synchrotron approaches can achieve from submicron to nanometric ($\sim$40 nm) spatial resolutions [17]: (a) parallel beam submicron tomography, (b) magnified tomography by projection, (c) magnified tomography using X-ray lenses and (d) coherent diffraction imaging techniques. X-ray lenses can be used as the equivalent of the objective lens of a microscope to magnify the image at the exit of the sample. In particular Fresnel zone plate (FZP) and to some extent compound refractive lens (CRL) based full-field microscopes are in operation at different synchrotron facilities. While particularly successful at very low X-ray energies, the poor high energy efficiency of the zone plates and the limited field of view (15–50 μm) make this approach less adapted to high energy studies in materials science. Coherent diffraction imaging techniques are proposed as an alternative route to achieve spatial resolutions far beyond what can be achieved with high resolution X-ray detectors or X-ray optics. Some material science applications have been reported [18], but today most materials science applications use either parallel beam submicron tomography (subject of Sect. 3) or magnified tomography by projection (subject of Sect. 4).

3 Parallel Beam Submicron Tomography

Submicron tomography at a synchrotron source can be achieved in parallel beam geometry. In this case, it should be reminded that no geometrical magnification can be obtained as the beam is not divergent. The geometry of a standard acquisition setup in the parallel configuration is shown in Fig. 2. It is clear from the figure that the key technology in acquiring tomographic images with submicron resolution lays mainly in the flux of the incident beam and, most of all, in the resolution and the pixel size of the imaging detector.

Let's first examine the flux problem: in parallel beam configuration for a given flux (number of photons reaching the detector area per second), the time required to collect a given number of photons scales with the square of the lateral pixel size of the detector. Decreasing the pixel size from 1 μm to about 0.3 μm implies, if nothing else is changed, of about 10 times longer counting time. To avoid these extremely long counting times, the solution is to increase the photon flux. This explains why submicron tomography is easier to apply at synchrotron sources where the high brilliance and flux of the beam are sufficient to perform tomographic scans within a reasonable time. This can for instance be achieved using multiple insertion devices in series and a "pink" rather than a monochromatic beam (multilayer monochromators rather than Si crystals).

Fig. 2 Geometry of the acquisition setup in a parallel beam configuration

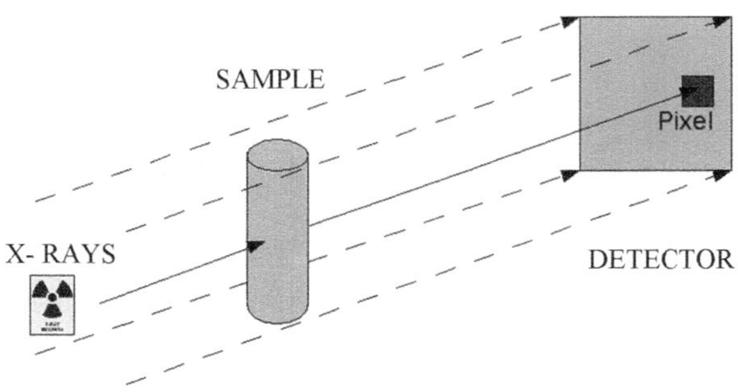

In terms of the quality of the imaging detector, imaging with submicron resolution is more delicate. The classical system for microtomography contains a scintillator transforming the X-rays into optical light that is imaged using classical optical microscope optics and coupled to a Charge Coupled Device (CCD). The technology of the optics, same as the one used in a standard optical microscope, can not really be improved and is in any case restricted to a resolution of about 0.3 μm because of the finite wavelength of the visible light. This ultimate resolution is in practice never reached in an X-ray detector as it would lead to an efficiency approaching zero (and an infinite acquisition time). The scintillator is the main component where one can gain. These have improved in recent years and some columnar scintillators do exist, but the influence of one pixel on its neighbors can not be avoided. When a pixel is irradiated by X-rays, it also influences its neighbors. This interaction effect and the optical lens coupling tend to blur the information over a finite neighborhood. This implies that the actual resolution (defined as the minimum distance between two features in order to separate them) is worse than the pixel size because of this mutual influence of the pixels on each other.

In summary, it is possible to design a parallel beam tomograph achieving a voxel size significantly smaller than 1 μm; the actual spatial resolution remains however in the range 0.7–1 μm at best. A system with a voxel size down to 0.17 μm has for instance been designed on the ID19 beamline at the European Synchrotron Radiation Facility.

4 Magnified Tomography by Projection

In order to overcome the spatial resolution limit of the detector, the image can be magnified by projection in a divergent beam. Figure 3 shows the experimental setup schematically. The geometric magnification M is given by:

$$M = \frac{z_s + z_d}{z_s}, \tag{2}$$

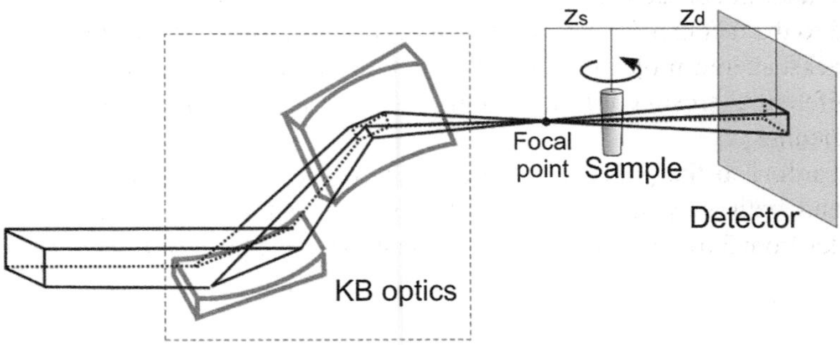

Fig. 3 Schematic diagram of the experimental setup of magnified synchrotron tomography experiments using KB optics at the nano-imaging end-station ID22NI. Reprinted figure with permission from [23], Copyright (2009) by ELSEVIER (Editor Subhash Mahajan)

where z_s is the distance between the focal point and the sample, and z_d between the sample and the detector. In this configuration a more efficient, coarser detector can be used and the spatial resolution is ultimately limited by the size of the focal point. To produce a focused and divergent beam at a synchrotron, multilayer coated mirrors are preferred because of their high efficiency also at higher energies, to the small focus that can be achieved and to the relatively large cone angle or numerical aperture. It should be noted that while this setup shows some similarities with the usual lab-based cone beam geometry, there are also important differences: (a) the brilliance is much higher leading to better spatial resolution and shorter exposure times and (b) the coherence and phase contrast effects become dominant. To obtain interpretable data at best spatial resolution (not limited by the interference fringes), a phase retrieval step must be associated to this configuration. This combined approach is called magnified holotomography. On the nano-imaging end-station ID22NI of the ESRF, the focal point is produced by a set of crossed multilayer coated bent mirror surfaces (so called Kirkpatrick-Baez optics or KB optics). They produce a beam with a size of 80 nm (H) by 130 nm (V) and medium monochromaticity ($\Delta E/E = 2 \times 10^{-2}$). The accessible X-ray energy range is 17–29 keV. Efficient phase retrieval for the holotomographic reconstruction is achieved from recordings at four distances, corresponding to four magnifications. A detector pixel size of 1–2.4 μm is used, with magnifications ranging from 2 to 50. A typical voxel size used in practice is 50 nm with a field of view of 75 μm. Most tomography scans are acquired in zoom or local tomography mode. A coarse scan of the complete sample is acquired, followed by a high resolution holotomography scan of a selected region of interest.

5 List of Synchrotron Beamlines with Tomographic Setups

Obtaining beamtime at a synchrotron source for tomography measurement is difficult. Therefore the knowledge of available instruments capable to image significant details of investigated structures using high energies is of great importance. Tables 1, 2 and 3 below summarize present state of the art at synchrotron sources around the world, which we hope that will help the potential user in obtaining easier the necessary beamtime and to perform an optimal measurement.

6 Examples of Parallel Beam Submicron Tomography

High resolution is required when the features of the microstructure to be analyzed have a very small size. An illustration of such a situation is the case of syntactic foams used in the oil industry to thermally insulate the pipes transporting the hot

Table 1 Overview of experimental setups for high energy *absorption* tomography using *parallel beam* (WB = white beam)

Facility/Beamline		Energy range (keV)	Min voxel size (μm)/ Max resolution (μm)	Min scan time at highest resolution
ESRF (FR)	ID19	7–60	$(0.3)^3/{\sim}1$	~15 min
	ID15 [37]	30–250^{WB}	$(1.1)^3/{\sim}2$	100 ms
	ID22	7–65	$(0.3)^3/{\sim}1$	~30 min
APS (USA)	2-BM-B [38]	5–30	$(0.67)^3/{\sim}1$	~6 s
	5-BM-C	10–45	$(2.4)^3/{\sim}4$	~6 h
	13-BM [39, 40]	6–70	$(1.0)^3/{\sim}2$	~30 min
	32-ID	8–35	$(0.3)^3/{\sim}1$	~250 ms
SPring-8 (JP)	BL20B2	8–113	$(2.74)^3/{\sim}1$	~100 min
	BL20XU [41]	8–113	$(0.2)^3/{\sim}1$	~25 min
	BL47XU [41]	6–37.7	$(0.2)^3/{\sim}1$	~25 min
DESY (DE)/HARWI-II [42, 43]		16–150	n.a./n.a.	n.a.
BESSY-II (DE)/BAMline [44]		6–80	$(1.4)^3/{\sim}4$	2–3 h
ANKA (DE)/TopoTomo [45]		6–35 < 40^{WB}	$(0.9)^3/{\sim}2.5$	~2.5 h
SLS (CH)/TOMCAT [46, 47]		8–45	$(0.37)^3/{\sim}1$	~10–15 min

oil from the extraction fields situated at a high depth in the cold sea water. While they prevent the oil to cool down, these materials also have to resist a 30 MPa hydrostatic pressure. They are composed of a polymer matrix surrounding hollow silica spheres reinforcing the porous structure and preventing it to collapse under the applied pressure. Due to their heterogeneous structure, these materials (composed of air, silica and polymer) are extremely difficult to observe using standard microscopy technique, but X-ray tomography has been successfully used for this purpose [19]. The determination of the thickness of the hollow spheres is very important for the understanding of the physical properties of this material. This thickness is typically in the 1–3 μm range. Thus, the most appropriate way to determine it has been to use the 0.28 μm pixel size available at the ESRF. An illustration of the type of microstructure obtained on these materials is shown in Fig. 4. It is composed of two reconstructed slices of the same sample at the same location but with the 0.7 μm (left image) and the 0.28 μm (right image) voxel size setups respectively (these voxel sizes correspond to physical resolutions of approximately 2 and 0.9 μm respectively, because of the mutual influence of the pixels as mentioned above). One can clearly notice that the thickness of the silica hollow spheres is very small and requires the best available spatial resolution. A drawback of the 0.28 voxel resolution is the very small field of view (0.56 mm). The example in the figure showing the same sample scanned at different resolutions indicates, however, the benefit of using a higher resolution. The typical thickness of the sphere walls is about 1.5 μm. This represents 2.5 pixels at high resolution and about 5 pixels at the highest resolution. In both cases a real quantitative measurement of the thinnest thicknesses is hazardous but it is clear from the picture that the smallest details in the microstructure (including also the pores in the polymer matrix) are more easily captured at higher resolution.

Table 2 Overview of experimental setups for high energy *phase contrast and holotomography* using *parallel beam*

Facility/Beamline		Energy range (keV)	Min voxel size (μm)/ Max resolution (μm)	Min scan time at highest resolution
ESRF (FR)	ID19 [48]	7–60	$(0.28)^3/\sim 1$	~15 min
	ID15 (white beam)	20–250	$(1.1)^3/\sim 2$	100 ms
	ID22 (propagation technique)	7–65	$(0.3)^3/\sim 1$	~30 min
APS (USA)	2-BM-B	5–30	$(0.67)^3/1$	~6 s
	13-BM (propagation technique)	6–70	$(1.0)^3/\sim 2$	~30 min
	32-ID	8–35	$(0.3)^3/\sim 1$	~10 s
SPring-8 (JP)	BL20B2 (Bonse-Hart interferometer)	15–25	$(11.7)^3/\sim 30$	~120 min
	BL20B2 (Talbot interferometer)	8–15	$(5.5)^3/\sim 12$	~90 min
	BL20B2 (propagation technique)	8–113	$(2.74)^3/\sim 10$	~100 min
	BL20XU (Bonse-Hart interferometer)	10–25	$(2.74)^3/\sim 10$	~180 min
	BL20XU (propagation technique)	8–37.7	$(0.2)^3/\sim 1$	~25 min
	BL47XU (propagation technique)	6–37.7	$(0.2)^3/\sim 1$	~25 min
BESSY II (DE)/BAMline [44]		6–80	$(1.4)^3/\sim 4$	2–3 h
ANKA (DE)/TopoTomo [45] (WB)		<40	$(0.9)^3/\sim 2.5$	~2.5 h
SLS (CH) [46, 47]	TOMCAT (propagation technique)	10–40	$(0.37)^3/\sim 1$	~10–15 min
	TOMCAT (differential phase contrast)	14–35	$(3.5)^3/\sim 5$	~10–45 min
ESRF (FR)	ID19 [48] (holotomography)	7–60	$(0.3)^3/\sim 1$	~15 min per distance
BESSY II/BAMline [44] (holotomography)		6–80	$(1.4)^3/\sim 4$	2–3 h per distance

Figure 5 shows another example of parallel beam tomography with submicrometer resolution. In this case, the material is an AlSi12CuMgNi alloy with 15 vol.% of Al_2O_3 short fibres [20] (hereunder AlSi12CuMgNi/Al_2O_3/15s). This is a metal matrix composite in which the reinforcement is formed by a highly interconnected hybrid 3D network of eutectic Si, ceramic short fibres and aluminides [21]. This 3D network provides high creep resistance to this composite in comparison with similar composites with discontinuous reinforcement [20–22].

The holotomographic technique has been used in order to reveal phases with similar attenuation coefficients such as the eutectic Si, the Al-matrix and the ceramic short fibres. For this, tomographic scans were carried out at the ID19

Table 3 Experimental setups for high energy *absorption* and *phase contrast* tomography using *magnifying optics*

Facility/Beamline		Type of contrast	Energy range (keV)	Max sample diameter size (μm)	Min voxel size/Max resolution	Min scan time (min)
Magnified synchrotron tomography using FZP						
SPring-8 (JP)	BL47XU [49, 50]	Absorption [51]	6–12	~70	$(40\ nm)^3/\sim(200\ nm)^3$ at 8 keV	~25
		Zernike Phase plate [52]	~8	~70	$(40\ nm)^3/\sim(200\ nm)^3$ at 8 keV	~25
		Differential phase contrast by Talbot interferometer	8–10	~70	$(40\ nm)^3/\sim(200\ nm)^3$ at 8 keV	~90
APS (USA)	26-ID (multilayer Laue lens [53])	Absorption/phase	8–10	~10	$(30\ nm)^3$/n.a.	120
	32-ID [54]	Absorption/phase	7–17	~25	$(11\ nm)^3/(40\ nm)^3$ at 8 keV	~20
SLS (CH)/TOMCAT		Absorption [55]	8–12	~50	$(16\ nm)^3/(144\ nm)^3$	~15–40
		Zernike phase contrast [56]	10	~50	$(16\ nm)^3/(144\ nm)^3$	~20
Magnified synchrotron tomography using Kirkpatrick-Baez mirrors						
ESRF/ID22 [23, 24]		Holotomography (e^- density map)	17–29	~400	$(50\ nm)^3/\sim(180\ nm)^3$	~80
Magnified synchrotron tomography using Compound Refractive Lenses						
ESRF/ID15A		Absorption	30–50	~100	$(100\ nm)^3/\sim(200\ nm)^3$	~5
Magnified synchrotron tomography using Bragg crystals						
BESSY II/BAMline		Absorption/phase	10–40	~100	$(150\ nm)^3/\sim1\ \mu m^3$	n.a.

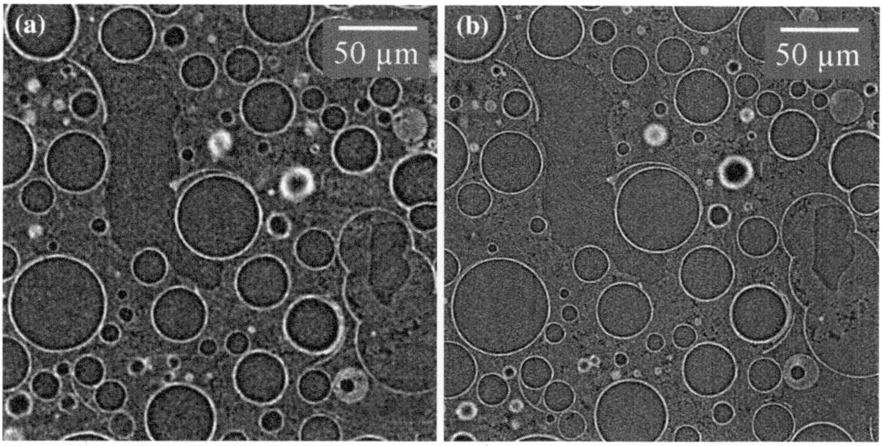

Fig. 4 Comparison of the same reconstructed tomographic slice (E = 20.5 keV, sample-to-detector distance 15 mm) of a polymer foam containing hollow silica spheres used in the oil industry as a thermal insulation in deep-sea water. The left picture (**a**) was recorded using 0.7 µm voxel size, whereas 0.28 µm voxel size was used for the picture on the right (**b**). The reconstructions show that the higher spatial resolution is required to image the very thin silica thickness of the reinforcement

beamline at the ESRF using beam energy of 20.5 keV. Scans at three different sample-to-detector distances were performed to achieve effective phase retrieval. Figure 5a, b are portions of reconstructed slices of the same region with voxel sizes of $(0.7\ \mu m)^3$ and $(0.28\ \mu m)^3$, respectively. Voids formed during the production of the composite (white), the ceramic short fibres (dark grey), the eutectic Si (light grey) and highly absorbing aluminides (black) are revealed embedded in a continuous Al-matrix for both resolutions. The zoomed regions in Fig. 5c, d show the higher level of detail obtained in the holotomographic reconstruction with higher resolution.

7 Examples of Magnified Tomography by Projection

For the following examples, magnified synchrotron tomography was carried out using beam energies up to 29 keV combined with a Kirkpatrick Baez (KB) optics system for Al and Ti-alloys as well as for Cu samples of a few hundred µm diameter [23]. Holotomography was applied to reveal microstructural features of phases that do not produce sufficient absorption contrast. Furthermore, the phase retrieval step correctly accounts for the Fresnel diffraction occurring in the projection geometry used. An X-ray microscope reported by Mokso et al. [24] has been used at the nano-imaging end-station ID22NI of the European Synchrotron

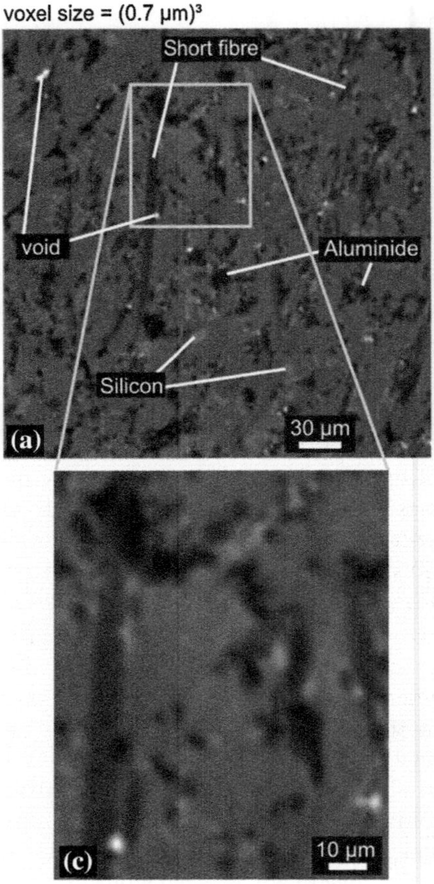

Fig. 5 Comparison of two holotomographic reconstructions of AlSi12CuMgNi/Al_2O_3/15s metal matrix composite (E = 20.5 keV): **a** voxel size = (0.7^3 μm^3), **b** voxel size = (0.28^3 μm^3). **c** and **d** show zoomed regions of (**a**) and (**b**), respectively

Radiation Facility (ESRF). 1200 projections of 1500 × 1500 pixels are acquired between 0 and 360° at 4 distances z_s = 29.68, 30.6, 34.6 and 44.61 mm using a CCD detector designed at ESRF [25]. The exposure time per frame is less than 1 s with a detector pixel size of 2.4 μm. The focus-to-detector distance ($z_s + z_d$) was kept fixed at 1185 mm and 1402 mm for the experiments at an X-ray energy of 17.5 keV (Al alloys) and 29 keV (Ti alloys and Cu), respectively. This results in minimum effective pixel sizes of 60 nm and 50.7 nm, respectively. Samples were machined to about 0.4 mm diameter for Al and Ti and to about 0.2 mm for Cu. The reconstructed volumes have a size of 1500 × 1500 × 1500 voxels with voxel sizes of 60 × 60 × 60 nm^3 and 50.7 × 50.7 × 50.7 nm^3, resulting in volumes of $(90\ \mu m)^3$ and $(76\ \mu m)^3$ for the Al-based and the Ti-based samples, respectively.

7.1 3D Microstructure of a Whisker Reinforced Titanium Alloy

Metal matrix composites (MMC) are obtained when a continuous metallic matrix is reinforced by means of a secondary stiffer phase, usually a ceramic [26]. The reinforcement of metals can have many different objectives such as an increase in specific mechanical properties like creep resistance, tensile strength, wear resistance and Young's modulus [26]. Accordingly, the requirements for the reinforcing phase/s can be low density, thermal stability, high strength, high elasticity modulus, etc. [27].

A Ti6Al4V alloy reinforced with 5 vol.% of TiB (hereunder called Ti64/TiB/5w) was produced at Crucible Materials Corporation by powder metallurgy. The raw components precursors of the Ti6Al4V alloy were mixed with 1 wt.% B and 0.1 wt.% C and this mixture was melted to produce powders by argon gas atomization [28]. The powder granules (<500 μm) are spherical with satellite particles and a mean particle size of 140 μm. Around 5 vol.% of TiB whiskers are formed during solidification. These powders were then HIPed at 1200°C for 2 h to obtain a dense material [29].

The 2D microstructure of the Ti64/TiB/5w composite revealed in SEM is shown in Fig. 6a where fine globular α grains (grey) and TiB whiskers with two size classes (black and dark grey) embedded in a β field (white) are visible. Figure 6b shows a portion of a reconstructed slice obtained by magnified holotomography using the KB optics. The α (light grey) and β (black) phases as well as the TiB whiskers (dark grey) can be clearly identified down to a size of about 180 nm.

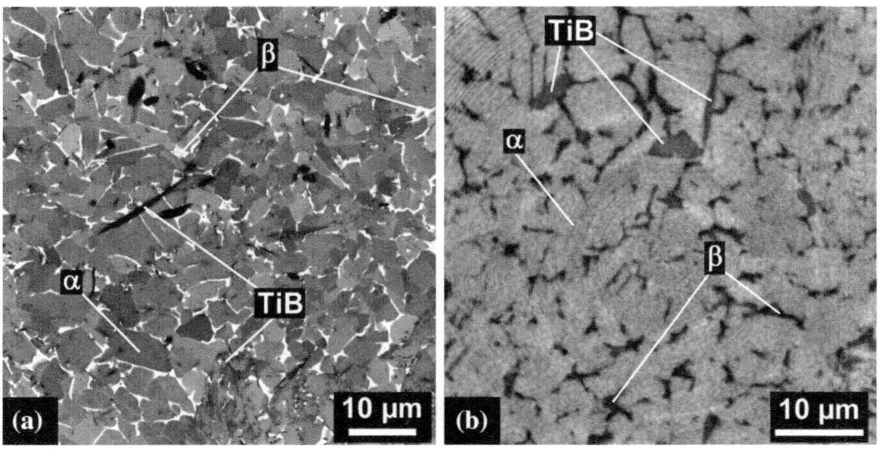

Fig. 6 2D microstructure of the Ti64/TiB/5w composite showing α and β phases and TiB needles revealed by (**a**) SEM in BSE mode and (**b**) magnified synchrotron tomography using KB optics

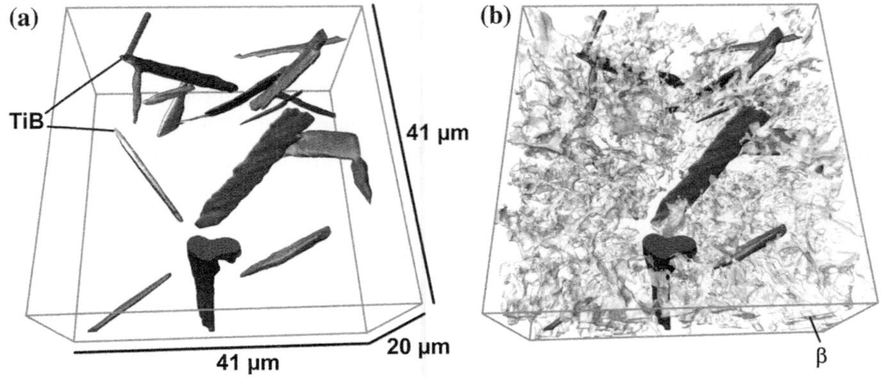

Fig. 7 Rendered volumes of the Ti64/TiB/5w obtained by magnified synchrotron holotomography using KB optics (voxel size = 50.7 × 50.7 × 50.7 nm^3): **a** TiB whiskers and **b** the same TiB whiskers with the largest β phase particle within this volume

Figure 7a shows a rendered volume (41 × 41 × 20 μm^3) of the Ti64/TiB/5w composite in which only the TiB whiskers of the larger size class are depicted. Their 3D morphology is clearly revealed showing that they can be present as individual or interconnected needles and elongated plates. The largest particle of β phase within the same volume is shown in Fig. 7b. The total volume fraction of β within the investigated volume is 9 vol.% and the volume fraction of this individual β particle amounts about 8.3 vol%. This phase has an irregular 3D structure with a high degree of interconnectivity and it is as well connected to the resolved larger TiB forming an interpenetrating structure within the α phase. In [30] the same material was investigated by parallel beam holotomography using a voxel size of $(0.3\ \mu m)^3$. The morphology of the TiB could not be resolved by this technique, but it was found that the TiB whiskers of the larger size class, such as those shown in Fig. 7, are distributed heterogeneously in the interspace of the prior powder granules. Analyses of hot deformed samples presented in the same investigations [30] show that these larger TiB whiskers are preferred damage location sites in the form of:

- fracture of TiB: this is promoted by the facts that the larger TiB needles amount only about 5 vol%, must support higher loads [30], they contain more defects than the smaller ones, and by stress concentration at the junctions between interconnected needles observed in Fig. 7a.
- debonding of matrix and TiB interfaces due to strain localization in the titanium matrix at the sharp borders of the TiB whiskers (see Fig. 7a).

These investigations show that the formation of these large TiB whiskers should be avoided during the atomization of the powders [30]. For this, faster cooling of the molten powders should be performed to promote more nucleation sites and to shorten the time available for in situ growing of the TiB [31].

Fig. 8 Scanning electron micrograph of the 2124/SiC/25p composite produced by ball milling and extrusion

7.2 Aluminium Alloy Reinforced with Multimodal SiC Particles

In [32] powders of a 2124 (AlCu4) Al-alloy were blended with 25 vol.% of SiC particles smaller than 5 μm by high energy ball milling (BM) using a planetary mill. The powder agglomerates are repeatedly deformed, fractured and cold welded during BM. The blended powders were then cold compacted in Al tubes and subsequently extruded. The resulting MMC (hereunder 2124/SiC/25p) is formed by an AlCu4 matrix and 25 vol.% of SiC particles with a size distribution ranging from about 5 μm (P95 percentile) down to 50 nm. Figure 8 shows a SEM picture of the 2124/SiC/25p composite where SiC particles with different sizes can be seen together with Cu- and Fe- rich aluminide particles typical for the matrix alloy [33].

The BM process results in a composite with significantly higher elevated-temperature strength and a better creep resistance than similar composites that are also produced by powder metallurgy but without the ball milling step [32]. This is a consequence of the broken SiC particles in the submicrometer range and the larger amount of oxides formed as the oxide skin of the Al powder is tore out during the BM process. The 3D microstructural features of this composite were explored by magnified synchrotron holotomography. Figure 9a shows a reconstructed slice obtained by this technique. The highly absorbing intermetallics (dark grey) and the SiC particles (grey) embedded in the Al-alloy matrix in a region of $60 \times 60\ \mu m^2$ are clearly resolved. The zoomed regions in Fig. 9b, c show that both aluminides and SiC particles in the submicrometer range are visible, whereas particle agglomerations are seen as "cloudy" regions. All the aluminide particles resolved (>180 nm) within one volume of $(15\ \mu m)^3$ are shown in Fig. 9d in black. In all cases, they are connected to larger SiC particles (shown semi transparent), a feature very difficult to verify in 2D micrographs.

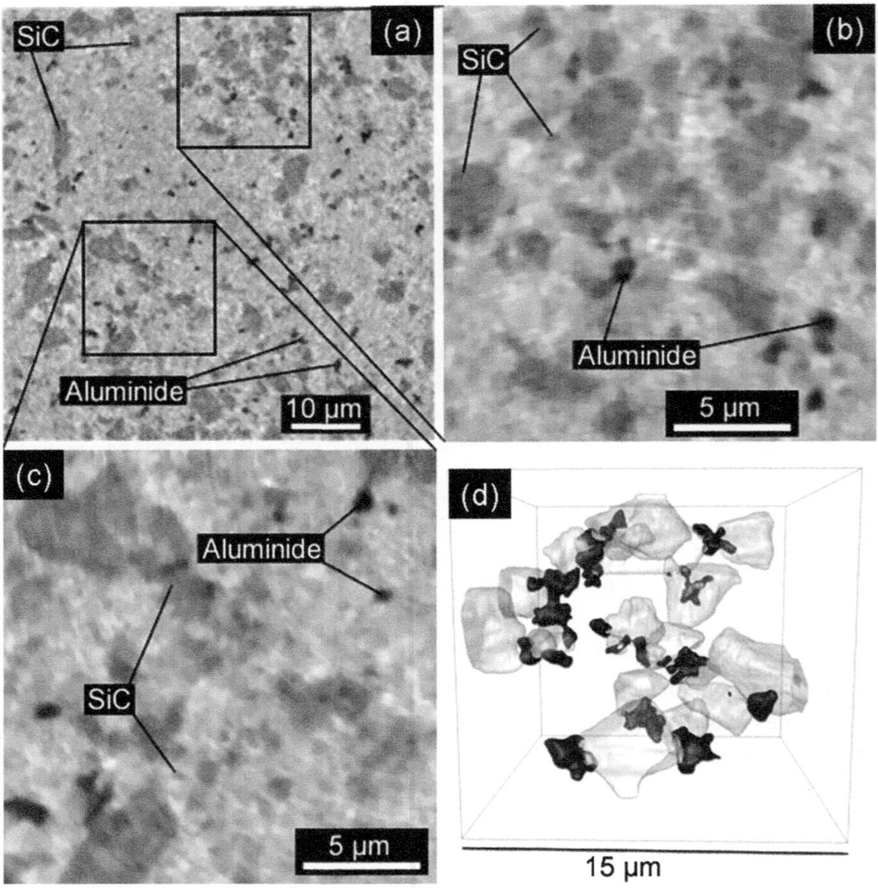

Fig. 9 **a** Reconstructed slice of the 2124/SiC/25p composite produced by ball milling and extrusion obtained by magnified synchrotron holotomography; **b** and **c** show zoomed regions of (**a**); **d** Rendered volume showing highly absorbing aluminides (*black*) connected to SiC particles (semi-transparent)

7.3 Shape Evaluation of Creep-Voids

Creep lifetime of metallic components subjected to mechanical load at high temperatures is limited by nucleation and growth of grain boundary voids. The shape of the voids is characteristic to the processes of nucleation and growth. For example energetic considerations predict a lenticular form for voids just after the nucleation [34], which may, however, change during the growth phase and becomes characteristic for the dominating growth mechanism. When this is diffusion (for creep at high temperature, small stress and small strain-rate) void shape can remain lenticular or change into needle like depending on the ratio between surface and grain boundary diffusion coefficients [35]. At deformation rates faster than the timescale of diffusive processes growth of voids is usually approximated as growth of a hole in a deforming nonlinearly viscous solid. In calculations a spherical void shape is generally assumed [36], which is however, difficult to identify with the real mechanisms of void growth by grain boundary sliding or by dislocation glide. Both processes are taking place at the submicron scale and lead to a distorted void shape.

Shape of grain boundary voids has been investigated in copper specimens creep deformed at constant stress of 12 MPa and temperature of 678 K. Magnified holotomography has been performed at the nano imaging end-station ID22NI of ESRF using a beam energy of 29 keV. The scanned specimens had a diameter of 0.2 mm and the different sample-detector distances were adjusted to obtain a voxel size of 50^3 nm^3. The change of voids shape due to plastic creep is evident from the comparison of Fig. 10a, b showing that the initially spherical (round) voids transform into a facetted (angular) form.

Submicron tomography turned out to be very useful in interpreting void growth data obtained in situ with fast-tomography during the power law creep process. The results indicate that void growth rate is proportional to void's volume but its value is much higher than the prediction of the continuum theory [36]. Since fast-tomography has a limited resolution (of about 2 μm) it was not clear if the found elevated growth rate is due to void coalescence or not. Submicron tomography reconstructions have shown that the average surface to surface distance between next-nearest voids is larger than the critical distance required for coalescence so that local mechanisms of dislocation glide/climb or grain boundary sliding should be responsible for the found elevated growth rates. This also means that continuum theory is inadequate to describe the growth of micrometer sized voids, which

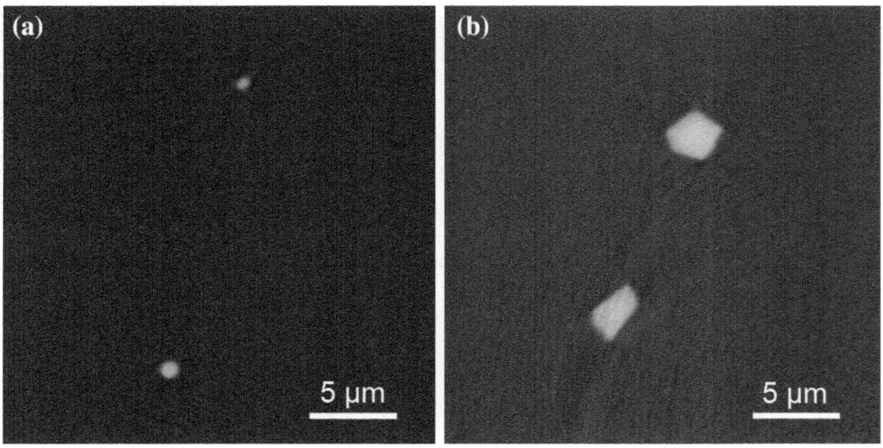

Fig. 10 Grain boundary voids (in *white*) in (**a**) recrystallized and (**b**) creep deformed copper (**b**). The shape of the voids changes under the action of plastic creep from a spherical (circular in 2D) into a facetted (angular) form

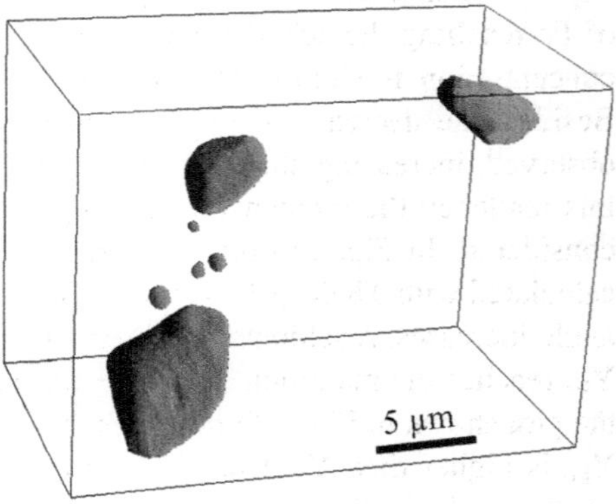

Fig. 11 Facetted grain boundary voids in creep deformed copper at 4% creep strain. Deformation temperature 678 K, applied stress 12 MPa

cannot be expected from a theory that does not take into account the crystallographic nature of the surrounding metallic matrix.

Submicron tomography has also revealed that lots of voids nucleate during creep, in agreement with the assumption of continuous void nucleation. Generally one grain boundary facet contains more than one void. The spatial distribution of large and newly nucleated small voids at 4 % creep strain is exemplified in Fig. 11.

Acknowledgments The electron microscopy images were obtained at the University Service for Transmission Electron Microscopy (USTEM) of the Vienna University of Technology.

References

1. Spanne, P., Rivers, L.: Computerized microtomography using synchrotron radiation from the NSLS. Nucl. Instrum. Meth. Phys. Res. B **24**(25), 1063–1067 (1987)
2. Flannery, B.P., et al.: Three-dimensional X-ray microtomography. Science **237**, 1439–1444 (1987)
3. Spanne, P., Throvert, J.F., et al.: Synchrotron computed microtomography of porous media: topology and transports. Phys. Rev. Lett. **73**, 2001–2004 (1994)
4. Borbély, A., Kenesei, P., Biermann, H.: Estimation of effective properties of particle reinforced metal-matrix composites from microtomographic reconstructions. Acta Mater. **54**, 2735–2744 (2006)
5. Stock, S.R.: Micro Computed Tomography: Methodology and Applications. CRC Press, Boca Raton (2008)
6. Banhart, J. (ed.): Advanced Tomographic Methods in Materials Research and Engineering, edited by. Oxford University Publishing, Oxford (2008)
7. Buffière, J.-Y., Maire, E., Cloetens, P., et al.: Characterisation of internal damage in a MMCp using X-ray synchrotron phase contrast microtomography. Acta Metall. **47**, 1613–1625 (1999)
8. Marrow, T.J., Buffiere, J.-Y., Withers, P.J., et al.: High resolution X-ray tomography of short fatigue crack nucleation in austempered ductile cast iron. Int. J. Fatigue **26**, 717–725 (2004)
9. Babout, L., Maire, E., Fougères, R.: Damage initiation in model metallic materials: X-ray tomography and modelling. Acta Mater. **52**, 2475–2487 (2004)

10. Maire, E., Bouaziz, O., Di Michiel, M., et al.: Initiation and growth of damage in a dual phase steel observed by X-ray microtomography. Acta Mater. **56**, 4954–4964 (2008)
11. Pyzalla, A., Camin, B., Buslaps, T., et al.: Simultaneous tomography and diffraction analysis of creep damage. Science **308**, 92–95 (2005)
12. Maire, E., Carmona, V., Courbon, J., et al.: Fast X-ray tomography and acoustic emission study of damage in metals during continuous tensile tests. Acta Mater. **55**, 6806–6815 (2007)
13. Cloetens, P., Ludwig, W., Baruchel, W., et al.: Holotomography: quantitative phase tomography with micrometer resolution using hard synchrotron radiation x rays. Appl. Phys. Lett. **75**, 2912–2914 (1999)
14. Bleuet, P., Lemelle, L., Tucoulou, R. et al.: 3D chemical imaging based on a third-generation synchrotron source. Trends Anal. Chem. **29**, 518–527 (2010)
15. Bleuet, P., Welcomme, E., Dooryhee, E., et al.: Probing the structure of heterogeneous diluted materials by diffraction tomography. Nat. Mater. **7**, 468–472 (2008)
16. Ludwig, L., King, A., Reischig, P., et al.: New opportunities for 3D materials science of polycrystalline materials at the micrometre lengthscale by combined use of X-ray diffraction and X-ray imaging. Mater. Sci. Eng. A **524**, 69–76 (2009)
17. Withers, P.J.: X-ray nanotomography. Mater. Today **10**, 26–34 (2007)
18. Barty, A., Marchesini, S., Chapman, H.N.: Three-dimensional coherent X-ray diffraction imaging of a ceramic nanofoam: determination of structural deformation mechanisms. Phys. Rev. Lett. **101**, 055501 (2008)
19. Adrien, J., Maire, E., Gimenez, N., et al.: Experimental study of the compression behavior of syntactic foams by in situ X-ray tomography. Acta Mater. **55**, 1667–1679 (2007)
20. Requena, G., Degischer, H.P.: Creep behavior of reinforced and short fibre reinforced AlSi12CuMgNi piston. Mat. Sci. Eng. A-Struct. **420**, 265–275 (2006)
21. Requena, G., Degischer, H.P., Marks, E., et al.: Microtomographic study of the evolution of microstructure during creep of an AlSi12CuMgNi alloy reinforced with Al2O3 short fibres. Mat. Sci. Eng. A **487**, 99–107 (2008)
22. Marks, E., Requena, G., Degischer, H.P., et al.: Microtomography and creep model-ling of a short fibre reinforced aluminium piston alloy. Adv. Eng. Mat. (2010). doi:10.1002/adem. 201000237
23. Requena, G., Cloetens, P., Altendorfer, W., et al.: Submicrometer synchrotron tomography using Kirkpatrick-Baez optics. Scripta Mater. **61**, 760–763 (2009)
24. Mokso, R., Cloetens, P., Maire, E., et al.: Nanoscale zoom tomo-graphy with hard x rays using Kirkpatrick-Baez optics. Appl. Phys. Lett. **90**, 144104 (2007)
25. Labiche, J.C., Mathon, O., Pascarelli, S., et al.: The fast readout low noise camera as a versatile x-ray detector for time resolved dispersive extended x-ray absorption fine structure and diffraction studies of dynamic problems in materials science, chemistry, and catalysis. Rev. Sci. Instrum. **78**, 091301 (2007)
26. http://mmc-assess.tuwien.ac.at
27. Kainer, U.: Basics of metal matrix composites. In: Kainer, U. (ed.) Metal Matrix Composites. Wiley-VCH, Weinheim (2006)
28. Yolton, C.T.: The pre-alloyed powder metallurgy of titanium with boron and carbon additions. JOM **56-5**, 56–59 (2004)
29. Poletti, C., Warchomicka, F., Degischer, H.P.: Local deformation of Ti6Al4V modified 1 wt% B and 0.1 wt% C. Mat. Sci. Eng. A-Struct. **527**, 1109–1116 (2010)
30. Poletti, C., Requena, G., Tolnai, D. et al.: Characterization of the microstructure and damage mechanisms in a Ti6Al4V alloy modified with 1wt%B. Int. J. Mater. Res. **101**, 1151–1157 (2010)
31. Banerjee, R., Collins, P.C., Genç, A., et al.: Direct laser deposition of in situ Ti-6Al-4V-TiB composites. Mat. Sci. Eng. A-Struct. **358**, 343–349 (2003)
32. Bauer, B., Requena, G., Lieblich, M.: Creep resistance depending on particle reinforce-ment size of Al-alloys produced by powder metallurgy. In: Proceedings of the 2009 International Conference on Powder Metallurgy & Particulate Materials (2009)

33. Davis, J.R.: Aluminium and Aluminium Alloys. ASM International, Specialty hand-book, Ohio (2004)
34. Riedel, H.: Fracture at High Temperatures, p. 70. Springer Verlag, Berlin (1987)
35. Chuang, T.J., Kagava, K.I., Rice, J.R., et al.: Non-equilibrium for diffusive cavitation of grain boundaries. Acta Metall. **27**, 265–284 (1973)
36. Budiansky, B., Hutchinson, J.W., Slutsky, S.: In: Hopkins, H.G., Sewell, M.J. (eds.) Mechanics of solids. The R. Hill 60th anniversary volume. Pergamon, Oxford (1982)
37. Di Michiel, M., Merino, J.M., Fernandez-Carreiras, D., et al.: Fast microtomography using high energy synchrotron radiation. Rev. Sci. Instrum. **76**, 1–7 (2005)
38. De Carlo, F., Xiao, X., Tieman, B.: X-ray tomography system, automation and remote access at beamline 2-BM of the advanced photon source. Proc. SPIE **6318**, 63180K (2006)
39. Rivers, M.L., Wang, Y., Uchida, T.: Microtomography at GeoSoilEnviroCARS. Proc. SPIE **5535**, 783–791 (2004)
40. Rivers, M.L., Wang, Y.: Recent developments in microtomography at GeoSoilEnviro-CARS. Proc. SPIE **6318**, 63180J (2006)
41. Uesugi, K., Takeuchi, A., Suzuki, Y.: High-definition high-throughput micro-tomography at SPring-8. J. Phys. Conf. Ser. **186**, 012050 (2009)
42. Haibel, A., Beckmann, F., Dose, T., et al.: The GKSS beamlines at PETRA III and DORIS III. Proc. SPIE **7078**, 70780Z (2008)
43. Beckmann, F., Herzen, J., Haibel, A., et al.: High density resolution in synchrotron-radiation-based attenuation-contrast microtomography. Proc. SPIE **7078**, 70781D (2008)
44. Rack, A., Zabler, S., Müller, B.R., et al.: High resolution synchrotron-based radiography and tomography using hard X-rays at the BAMline (BESSY II). Nucl. Instrum. Meth. A **586**, 327–344 (2008)
45. Rack, A., Weitkamp, T., Bauer Trabelsi, S.T., et al.: The micro-imaging station of the TopoTomo beamline at the ANKA synchrotron light source. Nucl. Instrum. Meth. B **267**, 1978–1988 (2009)
46. Stampanoni, M., Grosso, A., Isenegger, A., et al.: Trends in synchrotron-based tomographic imaging: the SLS experience. Proc. SPIE **6318**, 63180M (2006)
47. Marone, F., Hintermüller, C., McDonald, S., et al.: X-ray tomographic microscopy at TOMCAT. Proc. SPIE **7078**, 707822 (2008)
48. Cloetens, P., Barrett, R., Baruchel, J., et al.: Phase objects in synchrotron radiation hard x-ray imaging. J. Phys. D Appl. Phys. **29**, 133–146 (1996)
49. Toda, H., Uesugi, K., Takeuchi, A., et al.: Three-dimensional observation of nanoscopic precipitates in an aluminum alloy by microtomography with Fresnel zone plate optics. Appl. Phys. Lett. **89**, 143112 (2006)
50. Suzuki, Y., Toda, H., Ch, Schroer.: Tomography using magnifying optics. In: Banhart, J. (ed.) Advanced Tomographic Methods in Materials Research and Engineering. Oxford University Press, Oxford (2008)
51. Uesugi, K., Takeuchi, A., Suzuki, Y.: Development of micro-tomography system with Fresnel zone plate optics at SPring-8. Proc. SPIE **6318**, 63181F (2006)
52. Takeuchi, A., Uesugi, K., Suzuki, A.: Zernike phase-contrast X-ray microscope with pseudo-Kohler illumination generated by sectored (polygon) condenser plate. J. Phys. Conf. Ser. **186**, 012020 (2009)
53. Kang, H., Yan, H., Winarksi, R., et al.: Focusing of hard x-rays to 16 nanometers with a multilayer Laue lens. Appl. Phys. Lett. **92**, 221114 (2008)
54. Chu, Y.S., Yi, J.M., De Carlo, F., et al.: Hard-x-ray microscopy with Fresnel zone plates reaches 40 nm Rayleigh resolution. Appl. Phys. Lett. **92**, 103119 (2008)
55. Stampanoni, M., Marone, F., Mikuljan, G., et al.: Advanced X-ray diractive optics. J. Phys. Conf. Ser. **186**, 012018 (2009)
56. Stampanoni, M., Mokso, R., Marone, F., et al.: Phase-contrast tomography at the nanoscale using hard x-rays. Phys. Rev. B **81**, 140105 (2010)

Nano Characterization of Structures by Focused Ion Beam (FIB) Tomography

Flavio Andrés Soldera, Fernando Adrián Lasagni and Frank Mücklich

Abstract The 3D design of the micro and nanostructure of materials as well as the defects present in them, considerably determines the properties of modern materials. Therefore, a main focus in the current investigation and development trends is the direct influence of the structure of materials as well as its optimized design in continuously decreasing tolerances. However, such progress is only possible if adequate analysis techniques for the 3D representation of structures are present. The Focused Ion Beam (FIB) Tomography is a rather new technique, which allows the 3D reconstruction of structures in a range of scales and field of view which was previously, for instance through X-ray tomography or tomography in transmission electron microscopy, not available. Moreover, through the use of different signal detectors and the variety of materials that can be analyzed, it turns out to be a very versatile tool. The present chapter presents the working principles of FIB tomography, its different signal detection options, as well as different application examples. The very different examples related to aluminum alloys, multilayer coatings of

F. A. Soldera (✉) · F. Mücklich
Functional Materials, Saarland University, Campus D3.3, 66123 Saarbrücken, Germany
e-mail: f.soldera@matsci.uni-sb.de

F. Mücklich
e-mail: muecke@matsci.uni-sb.de

F. A. Lasagni
Materials and Processes department, Center for Advanced Aerospace Technologies – Andalusian Fundation for Aerospace Development (FADA-CATEC), Parque Tecnológico y Aeronáutico de Andalucia, C/Wilbur y Orville Wright, 17-19-21, 41309 La Rinconada (Sevilla), Spain
e-mail: flasagni@catec.aeroflasagni@us.es

F. A. Lasagni
MIM, Metalurgia e Ingeniería de los Materiales, Department of Materials and Mechanical Engineering, Escuela Superior de Ingenieros, Seville University, Camino de los Descubrimientos s/n., 41092 Sevilla, Spain

F. A. Lasagni and A. F. Lasagni (eds.), *Fabrication and Characterization in the Micro-Nano Range*, Advanced Structured Materials, 10,
DOI: 10.1007/978-3-642-17782-8_8,

different materials, oxidized nickel samples, porous materials and nanowires, intend to show the wide range of applications that these techniques can find as well as to awaken the imagination of the reader for thrilling new applications.

1 Introduction

Tomography methods can be divided mainly in two groups: projection techniques and serial sectioning techniques. Focused Ion Beam Tomography (FIB tomography) is a serial sectioning technique, where series of slices are milled in situ with a focused ion beam and then imaged by the ion beam itself or by an electron beam.

Acquisition of three-dimensional (3D) data by serial sectioning techniques consists of a controlled repetition of two-dimensional sectioning techniques and successive imaging or elemental mapping [1]. Such material removal techniques include dimpling, machining, polishing or microtomy. Recently developed processing automation has allowed for replacing the laborious and time-consuming practice of manual serial sectioning, significantly improving the speed up to $\sim$20 slices per hour. Besides the long time consumption, another drawback of these techniques is the relatively imprecise control of the distance between the slices. Distance between sections as small as 0.1 μm are reported in the literature [1] whereby the repeatability reaches 0.03 μm for sections of $\sim$0.8 μm. Typical values are however in the range of 1 μm. Depth profiling techniques, where layer-by-layer sample erosion is combined with laterally resolved surface analysis, can also deliver 3D data. One of the most versatile techniques is the sputter depth profiling, in which an energetic ion beam erodes the surface and a spectrometric technique like secondary ion mass spectroscopy (SIMS) or secondary neutrals mass spectroscopy (SNMS) characterizes the sputtered materials. Lateral resolution of 10 nm and depth resolutions in the nm range can be achieved with these techniques [2]. The main drawbacks are, on one side, that the ion impact may induce chemical damage on the surface to be measured and, on the other side, especially in heterogeneous materials, that the sputter ratio is difficult to control with high precision.

The recent development of the dual beam work stations having scanning electron microscopy combined with a focused ion beam (FIB) opened the possibilities for much more precise serial sectioning tomography. FIB tomography, also named in the literature as FIB microstructure tomography or FIB nanotomography, is a serial cutting technique where series of slices are worked out by the micromachining ability of the focused ion beam and are imaged by the ion beam itself or by an electron beam in its different contrast mechanisms. The ion beam is focused parallel to the surface to be milled minimizing possible damages to the original state of the surface. Moreover, this geometry allows for a much better control of the milled material so the sectioning can be achieved with a precision of some nanometers without prejudice to whether the material is homogeneous or not.

While the use of scanning electron microscopes in materials science originated already in the 1960s, the systematic development of liquid metal ion sources (LMIS) to be used for scanning ion microscopy took place in the 1980s [3, 4]. The driving force for the development and commercialization of FIB systems was given by the semiconductor industry, where they found applications in direct device fabrication, lithography instruments and semiconductor photomask repair tools, followed by circuit microsurgery to perform rapid prototyping or circuit diagnostics and failure analysis [5]. Sudraud et al. reported the use of a dual beam in 1988, consisting of a SEM with a gallium LMIS FIB column [6]. Commercial systems began to appear in the early 1990s [7] finding very soon novel applications in research laboratories.

First approaches of 3D analysis of materials using FIB were reported already in the 1980s, where serial sectioning was performed by milling slices with the FIB parallel to the surface to be analyzed [8, 9]. Since single beam FIBs were used, the samples had to be rotated after each milling and the milled surface was imaged with ions or a SIMS mapping was performed. However, limitations of the capabilities of computers were a main drawback for handling the generated data. In the 1990s further developments were made [10], however a decisive breakthrough was achieved with the use of dual beam systems (for instance Sakamoto et al. [11]), which allowed the improvement of the lateral resolution of the images by using electrons and opened the possibility of further contrast mechanisms, like for instance energy dispersive spectroscopy (EDS) or electron backscattered diffraction (EBSD) as will be discussed in this chapter.

2 Basics of FIB Tomography

FIB tomography consists mainly of two parts: serial sectioning with image acquisition, and image processing and 3D reconstruction.

The following procedures are typical for serial sectioning and image acquisition: selection of region of interest (ROI), in some cases in situ deposition of a protective layer on the ROI, positioning of the sample with its surface perpendicular to the ion beam, milling of trenches in front and at the sides of the ROI, parameters definition for tomography, and finally automated run of the slicing and viewing procedure. Of course the characteristics of all these steps depend on the sample to be analyzed, the resolution and field of view, the imaging signal to be used and the equipment characteristics.

Image processing and 3D reconstruction includes the alignment of images, in some cases the use of filters for improvement of contrast or sharpness, the 3D reconstruction and normally the segmentation of phases or features. After that, quantitative analysis may be applied for evaluation of results.

There is no special surface preparation for applying FIB tomography. It can also be applied to non-polished surfaces, as long as their roughness is smaller than the volume to be analyzed. The same requirements as for SEM apply: the sample has to be stable against vacuum and the electron beam, and it has to be conductive. The application on insulators is possible by coating the sample for instance with Au or C, but is often more difficult since charging effects may disturb the serial sectioning by FIB or the imaging with SEM. The majority of applications reported in the literature are for metallic samples. FIB tomography measurements of polymeric samples [12], polymer/clay nanocomposites [13], and biological samples [14] has also been performed, however are outside the scope of this chapter. In the present chapter we will provide examples of metallic and ceramic samples.

2.1 Preparation of Region of Interest

First, the region of interest of the sample's surface is identified by SEM imaging. A protective layer of Pt or other material like C or W is deposited over the area of investigation (Fig. 1a) using electron beam induced deposition (EBID). This in situ CVD deposition process consists on the addition of a precursor gas in the SEM chamber near the sample. The gas is adsorbed on the surface. Through the energy delivered by the electrons of the beam, the adsorbed gas decomposes in to non-volatile products, forming a deposition layer. Such process can also be produced by the ion beam (ion beam induced deposition—IBID) which normally enables a higher deposition rate than with the electron beam. However, the first atomic layers of the sample can be damaged (up to 20 nm in depth) if ions are used, especially when, before deposition, imaging with ions is required to find the region of interest.

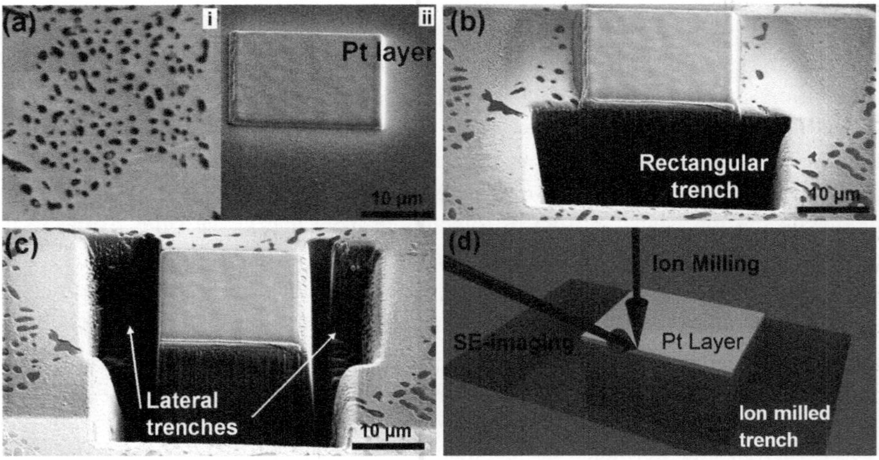

Fig. 1 Preparation of an Al sample for the FIB serial sectioning method: **a-i** I-beam micrograph (**a-ii**) deposited protective Pt layer on selected region of sample surface; **b** milling of the first rectangular trench; **c** milling of lateral trenches; and **d** schematic FIB cross section imaging methods indicating the direction of both the ion- and e-beams

The deposited layer is necessary to protect the sample against further working with ions, to suppress curtaining effects, thereby improving the polishing quality of the cross sections, and to obtain a sharp edge of the surface (between the sample surface and the Pt layer). This sharp edge will be necessary for the alignment of the images afterwards. If the sample surface (up to a depth of about 50 nm) is relevant for the investigations, a deposition with electrons is done first and then with ions. If the surface is not relevant, the Pt deposition can be directly done with ions.

Once the sample is tilted with its surface perpendicular to the ion beam and the region of interest is protected, material is removed (normally with the highest ion currents available) to produce trenches in front of the ROI, at the right and at the left sides (Fig. 1b, c). The trench in front of the ROI should have a thickness of about 1.5 times the depth of the region to be imaged. This trench is necessary in order to enable a good view, with the electron beam, of the cross section produced on the ions. The trenches at the sides should be large enough (about the same width as the depth) in order to avoid that redeposited material hinder the free view of the cross section. Moreover, since the secondary electrons detector is normally situated at the right side, the right trench avoids shadings effects in the deeper zones of the SEM images. The final schematic configuration after sample preparation is depicted in Fig. 1d.

A possibility to avoid the milling of a trench consists of selecting an area of interest near the edge of the sample (Fig. 2a). However, this configuration is only

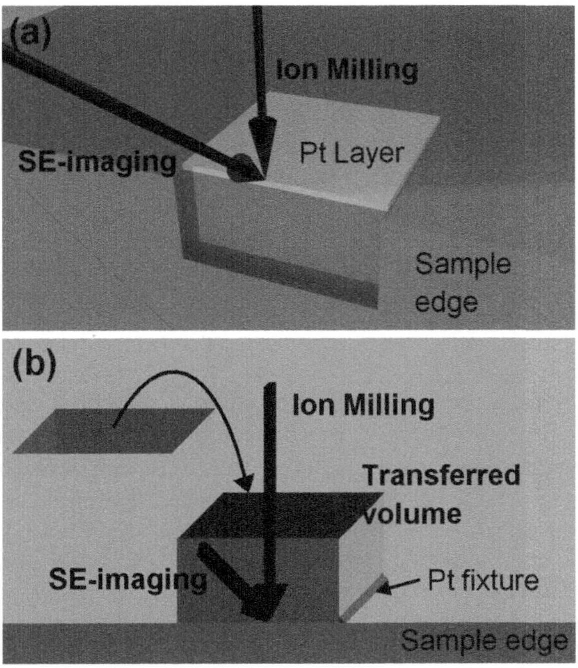

Fig. 2 Schematic FIB cross section imaging methods indicating both the direction of the ion- and e-beams **a** at the edge of a sample and **b** with the volume of interest transfer method. The angle between the electron and ion beam is 52° (for FEI equipment)

suitable for 3D characterization of samples where the morphological features of interest are found along such an edge or for the analysis of specimens with a homogeneous microstructure. Another option consists of cutting the volume of interest by ion milling and transferring it with a micromanipulator to the edge of the sample, fixing the volume using Pt deposition and performing the analysis there as described in [15] (see Fig. 2b).

2.2 Serial Sectioning

After removing enough material around the ROI, the cross section can be polished using a lower ion current. The current depends on the size of the region to be analyzed. For instance, if the area of the cross section is relatively large, e.g. 100 μm wide × 50 μm deep, the highest currents, about 9 nA, pA or 20 nA, may be used. For very small cross sections of about 2 μm × 2 μm, currents of about 50–100 pA should be used. Such currents will also be used for the serial sectioning.

Serial sectioning consists of the milling of thin slices of material with ions and the successive imaging with electrons. Normally secondary electrons are detected with an Everhardt–Thornley detector for large ROIs or a Trough-the-Lens or In-Lens-detector for relatively small ROIs. Other signals can be used, like imaging with ions, elemental mappings with EDS or crystallographic mappings using EBSD, as will be reported in Sect. 3 of this chapter.

The serial sectioning is done with a script which controls both electron and ion beams. In the case of equipment from FEI Company, the Auto Slice & View software is used. Typical milling for a slice ranges between 30 s and 5 min, depending on the size of the ROI and the ion current used. Imaging with electrons or ions may take between 20 and 90 s. Therefore, the serial sectioning process typically takes between 5 and 12 h, depending again on the size of the ROI.

2.3 Imaging

Besides the wide range of resolution and field of view of this technique, one of the main advantages of its application in materials science is the different imaging techniques that can be used. These imaging techniques are: secondary electrons (SE), backscattered electrons (BSE), electron backscattered diffraction (EBSD), energy dispersive spectroscopy (EDS), secondary electrons induced with ions and finally secondary ions. These imaging techniques provide different information from the samples, as will be shortly discussed in the following items. For more detailed information the reader should consult dedicated bibliography (e.g. Verhoeven [16]).

2.3.1 Imaging with Secondary Electrons (SE)

Secondary electrons are generated by the primary electron beam and also by scattered electrons that pass near the surface. The emission of SE is strongly concentrated in the entrance region of primary electrons, which allows for the best resolutions that can be achieved with the scanning electron microscope. SE have a relatively low energy, with a peak intensity at approximately 5 eV. Therefore, secondary electron detectors are biased to about 200 V to attract these electrons.

Image contrast with SE detector is produced mainly by the topography of the sample. Since the SE detectors also detects some backscattered electrons, whose emission strongly depends on the atomic number of the material, some material contrast is present. However it is quite small and much lower than that obtained with backscattered detectors.

Modern systems are nowadays equipped with two different SE detectors: the Everhart–Thornley detector, which has been the most widely used detector, and an in-lens detector situated above the objective lens so that the electrons spiral up to it. This configuration provides a higher resolution but is limited to non-ferromagnetic samples, since their magnetic field interacts with the objective lens field. Modern systems provide resolutions up to 0.9 nm if equipped with field emission guns.

Both detectors have been widely used for tomography. However, the in-lens detector has a lower depth of field; i.e. for large regions of interest it is not easy to keep the image in focus for all slices during milling in the z-direction. Consequently it is used for relatively smaller regions of interest ($\sim$2–5 μm^3), where the best resolutions are required.

2.3.2 Imaging with Backscattered Electrons (BSE)

Primary electrons that are scattered back out from the sample when they strike it are termed backscattered. Backscattered electrons have higher energies than secondary electrons and they shift their energy values as the primary beam energy is changed. Moreover, they are generated in a larger sample volume than secondary electrons; BSE images having therefore lower resolution. Besides topographical effects, the emission of BSE is strongly dependent on the atomic number of the material, electron channeling effects and magnetic domain effects, which allows for additional contrast mechanisms. Especially multiphase materials can be imaged even though no topography may be present.

BSE detectors generally have a ring geometry and are placed above the sample immediately below the plate of the objective lens. Dual beam systems have often room constraints, so that in some of them the imaging of cross sections with the sampled tilted perpendicular to the ion beam column, are not possible with BSE detectors. In this case, tomography is not possible and therefore few tomography experiments with BSE detectors are reported in the literature (e.g. [14]). An alternative is to change the biasing voltage of the SE detector to zero or negative values, in order to capture only backscattered electrons. However, the signal intensity is normally rather poor in this case.

2.3.3 Energy Dispersive Spectroscopy (EDS) Mapping

Energy dispersive spectroscopy (EDS) uses the characteristic X-rays emitted by the atoms as they are bombarded by electrons, to perform an elemental analysis. Compared with SE, X-rays are generated from a larger material volume, which leads to poorer lateral resolution.

A FIB consisting of both an ion and an electron source equipped with an EDS detector is capable of extending elemental analysis into the third dimension by sequentially milling the sample and analyzing each freshly produced surface [17, 18]. Fully 3D chemical analysis of materials can be directly performed using the EDS elemental maps obtained after each cross-sectional milling step. This procedure is ideal for materials that present a large number of phases and/or low SE/BSE contrast between phases. In [18], a fully automated slice-view-&-map routine was developed putting emphasis on improving stability during the automated slicing based on the geometry situation. Depending on the position of the EDS detector, EDS mapping can be done at the milling sample position or a movement of the sample may be required, which may lead to loss of precision.

Resolution in the EDS maps is material dependent. However, the pixel size is presumably smaller than the EDS resolution at the measuring conditions which ensure a high enough sampling of elemental data. Spatial resolution in the cutting direction depends on slice thickness. However, since the signal is generated at relatively large depths below the surface (some hundred nanometres), it makes no sense to make the slice thickness smaller than ~ 100 nm, if no kind of correction is attempted. Due to the relatively long time necessary for the EDS maps, a compromise between analyzed volume and number of slices must be found. Generally, slice thicknesses ranging between 300 and 500 nm permit the imaging of 6–10 μm long regions in the sliding direction at the expense of reduced resolution in the z-direction. Considering a time of approximately 30 min for each EDS map, about 10 h are necessary for the complete data collection of the sample, achieving a reasonably resolution. However, new EDS detectors like Si-drift detectors can reduce the mapping time by up to one order of magnitude through their larger count rates.

2.3.4 Electron Backscattered Diffraction (EBSD) Mapping

When backscattered electrons leave the sample, they may be diffracted by the crystal lattice. Electrons satisfying the brag condition with atomic planes will form characteristic lines in the phosphor detector, leading to the formation of a Kikuchi pattern or electron backscattered diffraction pattern (EBSP). By analyzing these patterns with the corresponding software, the crystallographic structure (phase) and its orientation with respect to a coordinate system can be identified for each measured point of the sample. The formation of EBSP is very surface sensitive, since those BSE leaving the sample are diffracted in the last 20 or 30 nm of the surface depth. This requires a very good sample preparation, since lattice distortions at the surface of the material will lead to bad qualities of the patterns and eventually to the inability for pattern indexing. For further information about EBSD, the reader is advised to consult specialized literature, for instance [19].

FIB polishing of the sample very often leads to good surface quality for the production of EBSP. Therefore, FIB milling with EBSD mappings can be combined to obtain 3D information of the sample, with the advantage that different phases can be identified and, even more importantly, quantitative information about the orientation of each measured point is available. Grains and grain boundaries may be easily represented and also misorientations maps can be produced.

The maximum intensity of EBSP is obtained when the sample is tilted 70° with respect to the electron beam. Therefore, the milling position and the mapping position for EBSD tomography cannot be the same. A tilting and/or a rotation are required depending on the geometrical configuration of the equipment. Through these movements between milling and imaging position, some loss of resolution cannot be avoided. Resolutions of $50 \times 50 \times 50\ \text{nm}^3$ have been reported in literature for relatively large analyzed volumes ($50 \times 50 \times 50\ \mu\text{m}^3$) [20], nevertheless it is expectable to reach better resolutions if smaller volumes are imaged.

Like the EDS mapping, EBSD also takes relatively long times compared with SE imaging. However, modern detectors and computing systems allow the measurement of up to some hundred points per second, so that a mapping can be done in several minutes. This depends very strongly on the quality of the sample polishing and consequently on the material to be imaged.

2.3.5 Imaging with Ion Induced Secondary Electrons (ISE) and Secondary Ion Mass Spectroscopy (SIMS) Mapping

The imaging capability of the ion beam can also be used for tomography. This was of particular interest at the beginning of FIB machine development, where single beam machines were most common. Ion beams emit secondary electrons from the surface, which can be detected with an Everhart–Thornley detector or also with a continuous dynode electron multiplier (CDEM) detector which also allows for detecting secondary ions. Since the sample is being sputtered during imaging, relatively small currents have to be used. Even though ion beams can be focused to about 5–7 nm, the resolution of such ISE images is determined by a combination of the primary ion beam diameter at the sample surface, the secondary electron/ion escape depth, and the implant depth/lateral spreading of the primary ions in the sample. Therefore, practical resolution may be on the order of 20 nm for most inorganic materials [10]. Ion beams provide a different contrast than electron beams. A decisive advantage of ion imaging is the improved channeling contrast, which results because the secondary electron yield varies as a function of crystallographic orientation within the sample. Thus, a single crystal region (i.e. a grain in a polycrystalline sample) will appear darker when it is aligned (or nearly aligned) to a low index direction due to a decrease in the number of emitted secondary electrons [21].

During tomography, the sample has to be tilted for changing from milling to imaging position. Tilting values of 90° [22] as well as 45° are reported [23]. Even in the same tomography, imaging with different tilt values (e.g. 30° and 45°) can provide additional contrast by taking advantage of the channeling contrast, as demonstrated by Inkson et al. for FeAl-based nanocomposites [23]. However, since the implementation of dual beam equipment, imaging with ions for tomography has decreased significantly and very few examples are found in literature. The main drawbacks are the necessity for sample movement as well as a poorer resolution than imaging with electrons.

In a similar way, instead of imaging with ion induced secondary electrons (ISE), a mapping with secondary ion mass spectroscopy (SIMS) can be done. Secondary ion mass spectroscopy (SIMS) is a very sensitive analytical technique that can provide parts per million (ppm) to parts per billion (ppb) sensitivity for most elements when relatively large areas of analysis are available. With the implementation of the FIB, a significant improvement of lateral resolution of SIMS has been achieved, at the expense of sensitivity. As the analytical volume is reduced, SIMS may not be able to provide the ppm to ppb sensitivity desired for trace identification, but can still provide good detection limits for matrix elements [24].

SIMS mapping requires longer dwell times than the ISE images, but chemical information like in EDS mapping is obtained. Lateral resolutions of up to 20 nm have been reported in the literature [22], which is significantly better than EDS mapping.

3 Application Examples

In this section, several application examples of metallic and ceramic materials are described. 3D reconstructions are obtained by using different acquisition methods (SE, EBSD and EDS). Samples are prepared using different geometrical configurations, as well.

3.1 FIB/SEM Tomography of Aluminum Alloys

These application examples report on recent investigations carried out in Sr modified $AlSi_7$ and $AlSi_8Mg_5$ alloys.

Al-Si alloys present a simple binary system with eutectic composition close to 12.6 wt% Si [25]. The microstructure depends on the Si content as well as on the casting process. Hypoeutectic alloys contain dendritic α-grains with the eutectic structure in between, formed by lamellar Si separated by the α-phase [17]. The Si solidifies during relatively fast solidification rates (i.e. squeeze casting) as interconnected Si networks of several thin lamellae ($\sim$1–2 μm) between the dendritic arms of primary α-Al [26]. The addition of Sr (0.1–0.3 wt%) modifies the eutectic morphology into a finer structure of a fibrous network [27] improving both ductility and tensile strength of the alloy [28].

Al–Si–Mg alloys with higher Si and Mg contents than in the wrought 6xxx series are gaining importance as foundry alloys by applying new casting technologies [29]. The ternary system is described in [30] as Al–Mg silicide alloys, since Mg_2Si segregates from the melt. Mg_2Si may solidify in complex shapes called "Chinese script" (according to 2D images). Different phases can be found in these alloys depending on the Mg and Si content, but also on the presence of impurities like Fe and Mn [31]. The control of the microstructure, for instance by semi-solid treatment, is essential for improving the mechanical properties for some applications [32]. To achieve this goal, it is extremely necessary to be able to represent the complex 3D geometry of the different phases at a very small scale, which can be properly done with FIB tomography, as is shown in the next examples.

Figure 3a depicts a reconstructed three-dimensional image obtained from 2D sectioning in $AlSi_7Sr$ [17]. The fibrous morphology of the modified eutectic is clearly observed, showing Si fibers and branches of about 0.2–0.4 μm in diameter and some micrometers in length ($\sim$3–5 μm) and oriented in many different directions. These fibers are mostly interconnected by nodes of some nanometers in cross section ($\sim$0.05–0.1 μm^2). The intersection of three different grains is observed in Fig. 3b separated by Si-free grain boundary zones about 1 μm wide. Grain A (see "raw data") shows a long thick Si fiber along the grain boundary with interconnected Si branches growing parallel to the *y*-axis [33]. Complete interconnection and similar orientation of fibrous Si is observed in grain B, while Si fibers in grain C are oriented perpendicular to those of grain A and B. In this grain (C), Si fibers seem to be mostly interconnected but some distance from the grain boundary, isolated short fibers are also observed.

Due to the small size of Si-features in the Sr modified alloy, FIB tomography appears as the most suitable technique for three dimensional characterization. Other techniques like micro or synchrotron X-ray tomography would not provide enough resolution for the reconstruction of the Si phase, also presenting a low phase contrast between Al and Si. Other sectioning methods would also not ensure a sufficient resolution in the slicing direction.

The ability to characterize the shape of the eutectic Si fibers (surface and volume) also depends on image processing parameters such as the use of smoothing algorithms [33]. As an example, using the software package Amira®, different smoothing options were employed. In Fig. 3b the 3D reconstructed images using constrained and unconstrained smoothing filters are presented. These options generate sub-voxel weights, such that the surface is naturally smoothed (a detailed description of the smoothing options is described in [34]). The lowest values for both the Si surface and volume, as well as the smallest surface to volume ratio (9.8 μm^{-1}) are observed by using the unconstrained render option.

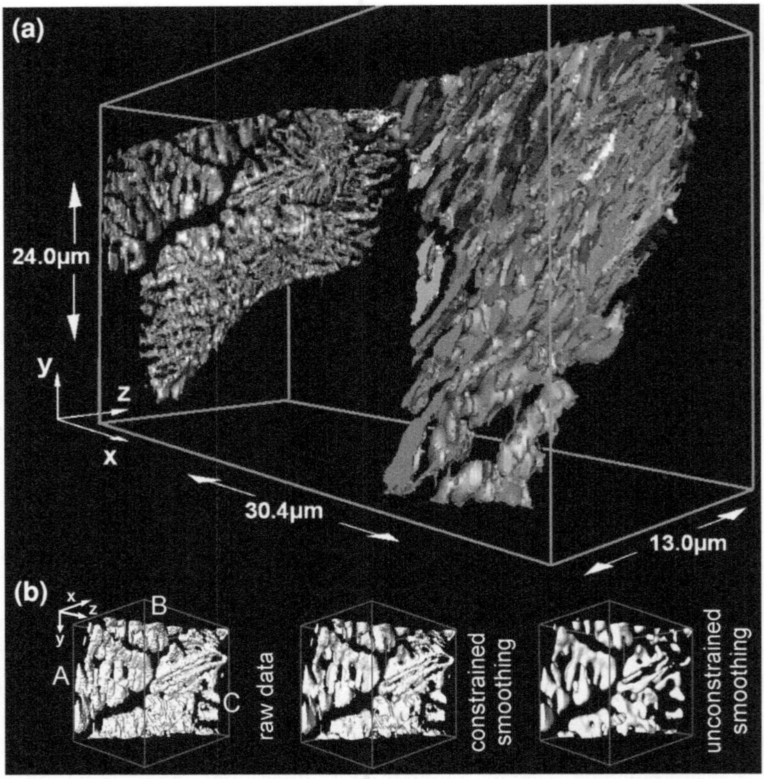

Fig. 3 3D reconstruction of eutectic Si branches in $AlSi_7(Sr)$: **a** reconstruction of the whole analyzed volume and **b** representation views of raw data, constrained and unconstrained smoothing render options. Voxel size (nm): $x = 60$, $y = 75$, $z = 60$ (slicing direction)

The important reduction of 34% in volume (when compared with raw data) produces a high distortion of the analyzed data generating inadequate 3D representation and quantification of the structure morphology. Moreover, after the rendering operation small particles are completely removed. On the other hand, while the surface/volume aspect ratio is equal for both raw and constrained options, a reduction in volume of around 20% is observed using the constrained smoothed render. This effect is due to the smoothed surface of the Si fibers.

In summary, special attention must be given for the calculation of microstructural parameters. Results obtained from raw data would provide large surface values but a more accurate volume. For the case of small features (in the range of tens of the voxel resolution), 3D tomography extracted from raw data would assure the representation of the different material constituents since the rendering options studied in this example can lose the recorded 3D information due to interpolation issues. For larger features, the constrained smoothing option looks like the best reconstruction option. Although the volume of the analyzed features will be slightly decreased, a better quantification of the surface would be achieved.

Figure 4a, b shows the 3D reconstruction of a Mg_2Si particle extracted from a tomograph of a gravity cast $AlSi_8Mg_5$ sample, which consists of a central node from which arms of surfaces emanate in radial directions [35]. In Fig. 4b, the different Mg_2Si lamellae seem to perfectly enclose the α-Al dendrite, from where other plates start to growth. It is very interesting to observe that the lamellae are completely interconnected, which was not the information obtained with 2D images (Fig. 4c). Moreover, quantitative information can be obtained. The total volume of the lamellae amounts to 2,490 μm^3 while its surface is 7,700 μm^2.

Shape characterization of the different features can be performed by different shape factors like circularity (C_P) and sphericity (S_P) for 2D and 3D characterization. These factors [27] are defined as:

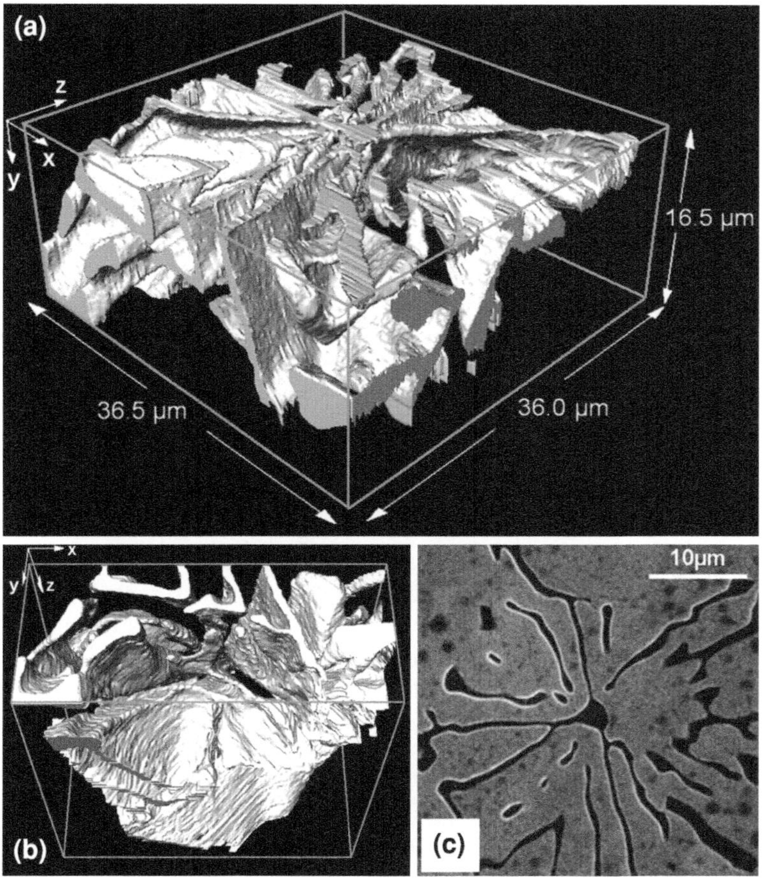

Fig. 4 FIB/SEM analysis of Mg_2Si Chinese script: **a**, **b** 3D representation views of the selected volume obtained with FIB tomography in sample 1; **c** SE-image obtained from the top of the region of interest before sample milling [31]. Voxel size (nm): $x = 42$, $y = 54$, $z = 200$ (slicing direction)

$$C_P = \frac{4\pi A}{P^2} \tag{1}$$

$$S_P = \frac{S^{3/2}}{6\pi^{1/2} V} \tag{2}$$

where, A, P, correspond to the area and perimeter in 2D; and V, S to the volume and surface area in 3D, respectively. Sphericity and circularity values are limited between 0 and 1. A shape factor of 1 corresponds to a perfect circle or sphere and decreases the more non-convex and elongated the particles are.

For the case of the Mg_2Si structure, the 3D sphericity factor is around 0.04, while the 2D circularity amounts to 0.2 with a standard deviation around the mean value. The latter exceeds the 3D value by 4.5 times. Only the biggest particle (56 μm^2 in area and ~35% of the Mg_2Si phase in the 2D micrograph) shows a C_{2D} factor of 0.02 comparable to that of the 3D reconstruction, while it increases with reducing particle area (from ~0.1 to ~0.7 for particle volumes of 18 and 3 μm^2, respectively).

The microstructure growth mechanisms can be perfectly analyzed with the 3D representations. Figure 5a, b shows a 3D reconstruction of a semi-solid heat treated sample. Three different sections containing eutectic Si phase are depicted in blue, red and green. Each colored region represents a zone of eutectic fibrous structure which has solidified with preferential orientation. The Mg_2Si phase (yellow) presents a central plate (oriented in the yz plane of the tomograph) from where several lamellae emanate perpendicularly (xy plane). This phase looks much thinner than the corresponding one from the gravity cast sample (Fig. 4). Figure 5c shows a SE-cross section containing both red and blue regions, which are represented in 3D by Fig. 5d, e, respectively. The Si fibers of the red grain are parallel to the xy plane with an orientation of around 30° with respect to the x-axis. In case of fibers in the blue grain, they are oriented following the z-axis.

In this application example, FIB-SEM tomography appears as the most suitable reconstruction technique. From one side, it allows the 3D characterization of the relatively large Mg_2Si "Chinese script" and the reconstruction of the Al–Si eutectic phase. Secondly, detailed information on the Si fibers orientation in the semi-solid modified eutectic structure can be obtained, even for 50–150 nm diameter branches which are not detectable even by high energy synchrotron tomography. Furthermore, both analyses were performed in the same tomography.

3.2 FIB/SEM Tomography of CVD Multilayer Coating

Wear resistance of cutting tools is usually improved by coating them with thin films. Depending on the coating type and application, the thin films are produced by chemical vapor deposition (CVD) or physical vapor deposition (PVD). For metal

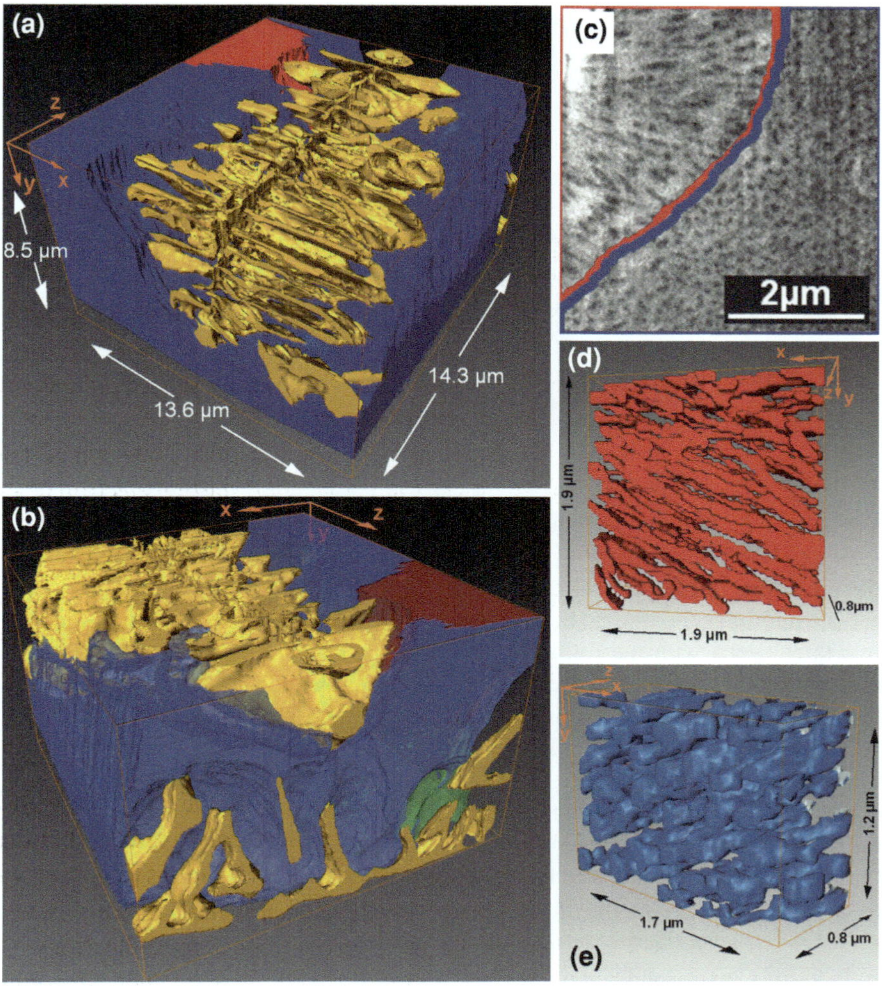

Fig. 5 3D reconstruction of semi-solid $AlSi_8Mg_5$ alloy showing: **a**, **b** Mg_2Si Chinese script (*yellow*) and regions containing Al–Si phases with preferential Si orientation (*red*, *blue* and *green*); **c** SE image of the cross section showing Si (*dark*) and Al (*light*) in red and blue regions; 3D Si reconstruction of regions **d** red and **e** blue [35]. Voxel size (nm): $x = 18$, $y = 23$, $z = 75$ (slicing direction)

turning cutting applications, multilayer coatings made by CVD are state-of-the-art. Typical wear resistant materials deposited by CVD are TiN, Ti(C,N), Zr(C,N) and Al_2O_3. The multilayer sequence of the coating aims to provide wear, corrosion and heat resistance [36]. Pitonak et al. have recently developed novel Ti(C,N) coatings with a gradient in grain morphology and composition, which were produced by modifying the deposition conditions during chemical vapor deposition [37]. The improved wear resistance is provided by a modified microstructure of the

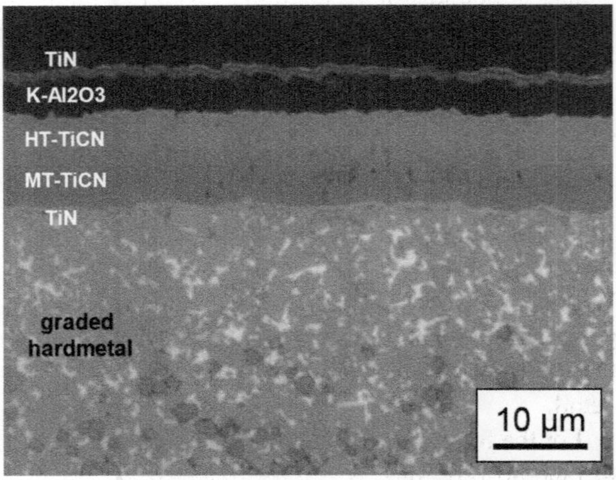

Fig. 6 Light microscopy pictures of CVD multilayer coating: the novel multilayer consists of a TiN/MT-Ti(C,N)/HT-Ti(C,N)/κ-Al_2O_3/TiN thin film sequence [34]

Ti(C,N) phase. The spatial architecture of the interface between the Ti(C,N) and the κ-Al_2O_3 layer, as well as of the Ti(C,N) grains should play a major role in the properties of the insert and should be studied by FIB tomography.

The multilayer coating to be analyzed is shown in Fig. 6. The coating is made of five thin layers, with the following sequence: a TiN thin layer of 0.5 μm which ensures a good adherence to the cemented carbide substrate, a MT-Ti(C,N) of 5 μm thickness, a HT-Ti(C,N) of 5 μm, a 4 μm κ-Al_2O_3 and a 0.5 μm TiN top-layer [38]. According to Fig. 6, the HT-Ti(C,N) layer shows as a dark gray colored region (dark-orange and light-orange in original light optical metallograph [34]), which corresponds to a nitrogen-rich carbonitride, as was previously verified by X-ray diffraction investigations determining a C/N ratio of 0.16/0.84 for the HT-Ti(C,N).

The FIB tomography technique was used in order to investigate the 3D arrangement of the grains in the transition between the HT-Ti(C,N) and the κ-Al_2O_3 top-layer. Series of cross sections with FIB and images with FE-SEM were performed perpendicular to the layers. A total volume of $22 \times 23 \times 12\ \mu m^3$, with a distance of 50 nm between cuts, was analyzed. Phases and some grains were segmented using the software Amira®. The images from the slices were affected in some cases by a strong curtaining effect, which would mainly interfere with the HT-Ti(C,N) grains segmentation [38]. To overcome this issue, periodic and symmetric patterns were removed from the frequency spectrum after applying the Fast Fourier Transform (FFT) to each slice with the software A4i. Afterwards, with the Inverse FFT, the spectrums were reversed to the original images but with almost no remaining traces of the curtaining effect.

Given the high contrast of the WC phase in the cemented carbide substrate and the κ-Al_2O_3 top layer respect to the Ti(C,N), the segmentation of these phases could be performed completely by gray scale thresholds. The Ti(C,N) layer was the most problematic area in terms of segmentation capability. Despite the many grains present, being a single phase material it resulted in a very low contrast among them. Individualization of complete grains required meticulous visual

Fig. 7 3D FIB tomography of novel multilayer system, showing WC grains of cemented carbide, the morphology and distribution of selected HT-Ti(C,N) grains and the κ-Al_2O_3 top-surface layer. Voxel size (nm): $x = 30$, $y = 38$, $z = 50$ (slicing direction)

inspection of the slices because, even when a relatively good contrast is devised, it may happen, that in some slices the contrast between the grain and the surrounding area is hard to distinguish. This makes the grain unsuitable for analysis. As for the suitable ones, each of them had to be differently processed: smoothing, edge detection and shading correction filters, contrast/brightness manipulation and morphological operations were selectively employed to enhance the visualization of each grain, leading to their manual segmentation.

The final 3D reconstruction is shown in Fig. 7. The dark red substrate represents the WC grains of the cemented carbide and the purple top-layer is the alumina coating. Between them, several HT-Ti(C,N) grains are reconstructed in different colors. The Ti(C,N) grains present a particular architecture, which is very different than the well-known equiaxed grains of High Temperature-CVD coatings or the columnar grains of Medium Temperature-CVD Ti(C,N) layers. The novel HT-Ti(C,N) layers present a preferred texture in the direction $\langle 110 \rangle$, as determined by XRD [39]. This texture corresponds with the observed $\langle 110 \rangle$ texture for star-shaped crystallites on TiN coatings. Nevertheless, some other crystallites with a lenticular-like morphology having a $\langle 211 \rangle$ texture are present in the layer microstructure, providing a mixed morphology in the layer. During the modified CVD deposition star-shaped Ti(C,N) grains with $\langle 110 \rangle$ texture form in the HT-Ti(C,N) layer and they prevail in the competitive growth with the lenticular-like crystallites, which form at higher deposition temperatures. From the 3D reconstruction, some preferred crystallite growth in a direction perpendicular to the substrate is observed. However, the Ti(C,N) crystallites present a non-typical morphology, with some grains showing screw-like type morphologies. Here it may be pointed out that the reconstruction intended to follow the pattern of a particular grain for the 3D representation of one single Ti(C,N) grain.

Moreover, results of FIB tomography confirm that Ti(C,N) crystallites at the interface between the HT-Ti(C,N) and the κ-Al_2O_3 layers present a needle-like structure. These Ti(C,N) protruding crystallites penetrate into the κ-Al_2O_3 top-layer, which provides mechanical anchoring between the layers.

3.3 FIB/EDS Tomography of Squeeze Cast $AlSi_{12}$ Alloy

In several applications it is not trivial to distinguish the different phases present in the microstructure from the SE images. Using EDS mapping can help in differentiating them. The disadvantage is that the separation of the slices cannot be as small as with SE images, since the depth of information of the EDS signal is some hundred nanometers. Also the lateral resolution is reduced. In this example the ability for reconstructing 3D information in an $AlSi_{12}$ alloy is presented (although the alloy constituents can even be characterized in the SE mode).

Figure 8a shows the reconstructed 3D volume of an $AlSi_{12}$ alloy obtained from 20 EDS cross sections with a separation of 310 nm and viewed from the orthogonal directions. Several interconnected thin Si plates are recorded showing different orientations amounting to $\sim$14% in volume fraction. The individual thickness of the Si lamellae are around 0.5–1 μm. Due to the large spacing between each cross section, the rendering option "constrained smoothing" [34] is not able to generate a smoothed surface of the Si plates in the z-axis (aliasing effect [40]) particularly at particles that show a pronounced oblique orientation with respect to the zx-planes. In this case, the surface and volume of the lamellae may be overestimated, leading to wrong morphological quantities. An example of the aliasing effect is shown in Fig. 8b, where a crop region of $10.8 \times 7.4 \times 7.1\ \mu m^3$ is shown.

A typical EDS map showing the Si content in the cross section of the same alloy is depicted in Fig. 8c. Considering a time of approximately 30 min for each EDS map, about 10 h are necessary for the complete data collection of the sample. New detectors like Si-drift detectors could however reduce the mapping time to a few minutes, considerably reducing the total measuring time. The shadowing effect can be very well appreciated from Fig. 8c. It results from the position of the EDS detector and the geometry of the milled rectangular trench, which in this case was not large enough.

3.4 FIB/SEM–EDS Tomography of Gravity Cast $AlSi_8Mg_5$ Alloy

In the case of multiphase materials, the different grey values of the observed features recorded by SE imaging may not be possible to be assigned to a specific phase since they are usually overlapped.

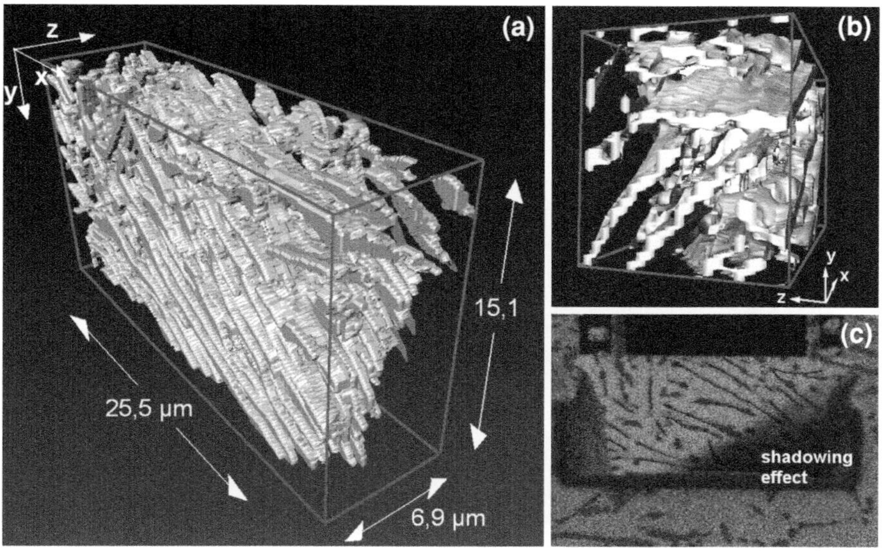

Fig. 8 **a** 3D reconstruction from EDS maps of squeeze cast $AlSi_{12}$ alloy; **b** cropped region ($10.8 \times 7.4 \times 7.1\ \mu m^3$) showing the aliasing effect due to the large distance between serial slices (310 nm) in z direction (slicing axis); **c** EDS map of an $AlSi_{12}$ cross section (dark = Si; bright = Al) denoting the shadowing effect due to the EDS detector position. Voxel size (nm): $x \sim 100$, $y \sim 100$, $z = 310$ (slicing direction)

As an example, a SE-image of $AlSi_8Mg_5$ alloy and its equivalent elemental maps (for Al, Si, Mg and Fe) are presented in Fig. 9a, b, respectively. Here, the identification of the different phases can only be performed with the help of an EDS detector. Four different grey contrasts are observable in the Al phase: Different grains (or subgrains) "a" and "b" present a light grey contrast compared to "c" and "d". The AlFeSiMg particle at the upper-right corner looks bright in Fig. 9a, and is identified in Fig. 9b. The aluminum phase in grain "c" presents similar grey values as the Mg_2Si phase, where only particle contours can be identified by SE imaging. Due to shadowing effects, grain "d" depicts a wide grey range which includes that of the Mg_2Si and AlFeSiMg phases. Only the Si phase (features in dark grey) is clearly recognizable in the whole studied image.

An optimized 3D reconstruction can be obtained by combining SE images and EDS mappings: SE images are used for drawing the phase contours (achieving higher xy-voxel resolution if compared with FIB/EDS tomography). One EDS map, like the one in Fig. 9b, is used every five SE-images for phase identification. Furthermore, the voxel resolution in the milling direction is improved with respect to FIB/EDS tomography since SE imaging and low resolution EDS maps consume considerably less time: 100 SEI + 20 EDS maps can be obtained in $\sim$10 h for cross sections of about $40 \times 20\ \mu m^2$ with a

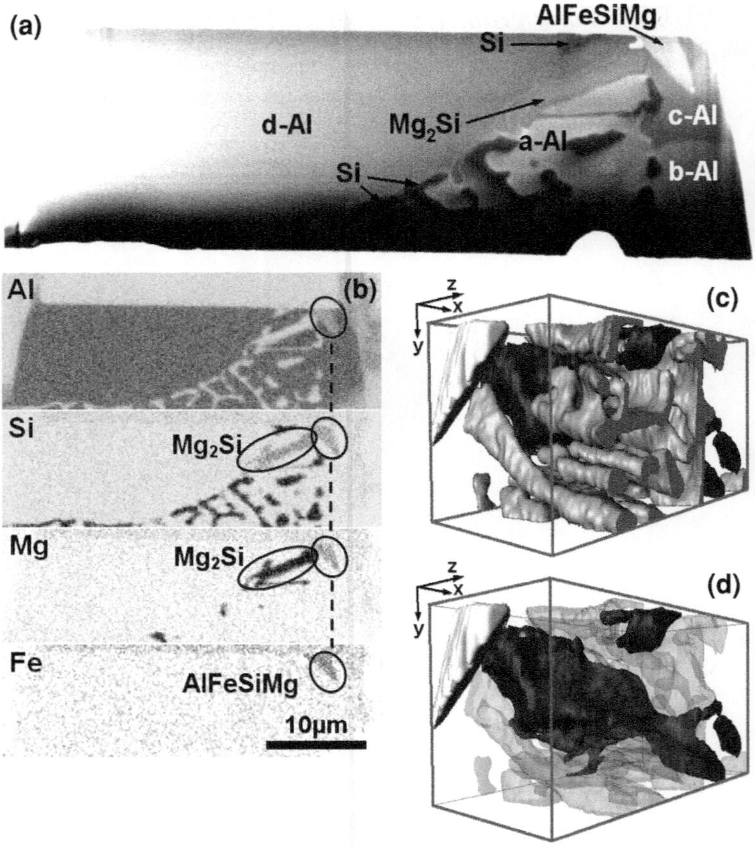

Fig. 9 **a** Secondary electron image (SEI) showing different phases (as indicated) and four different Al grains labeled *a–d*; **b** corresponding EDS elemental maps of a FIB cross section ($AlSi_8Mg_5$ alloy); **c**, **d** 3D reconstruction of the investigated alloy showing AlFeSiMg aluminide (*white*), Mg_2Si Chinese script (*light grey*), and Si lamellae (*dark grey*). The Si phase in **d** is depicted as semitransparent for better visualization of the Mg_2Si Chinese script [33]

resolution in the milling direction of 100 nm. The reconstructed volume of the $AlSi_8Mg_5$ alloy is depicted in Fig. 9c with transparent Al. A lamellar Si structure (11.5% in volume fraction) connected to the Mg_2Si phase (dark grey) is observed forming an interpenetrating network. The Mg_2Si phase amounts to ~7% in volume fraction for the reconstructed volume, showing a segment of the so-called "Chinese script" structure (from 2D metallographs) as observed in Fig. 9d, where the Si phase is represented as semitransparent. A large AlFeSiMg aluminide nucleated at the Mg_2Si particle is partially observable in the recorded volume, presenting a needle like structure.

3.5 FIB/EBSD Tomography of Craters Produced by Electrical Discharges in Oxidized Nickel Samples

Nickel is used as electrode material in spark plugs as well as in contact materials. Since in the case of spark plugs the oxidation plays an important role, oxidized Ni samples were tested under the action of electrical discharges similar to those in combustion motors. Such a discharge generates one or more craters in the material surface [41], breaking the oxide layer and producing some melting and resolidification. The purpose of the study was to evaluate, in three dimensions, the microstructural changes produced in the material by the energy input of the discharge.

The Ni sample was polished with diamond particles down to a grain size of 0.5 μm. Oxidation of the sample was carried out in air at 900 °C during 30 min (thickness of the oxide scale was approximately 5 μm). Individual sparks were produced between the cylindrical sample acting as the cathode and a platinum electrode (anode) with a constant separation distance between them of 1 mm. The discharges were produced in nitrogen at 7 bar.

In order to understand the physical processes involved in the development of the plasma erosion craters, a complete understanding of the 3D structure of the crater is crucially important. An open question is if there is any microstructural change in the different oxide layers or in the metal underneath the crater in addition to material loss by evaporation or particle ejection. For a first overview a serial cross sectioning was performed through an erosion crater (Fig. 10a, b). Serial cross sectioning of plasma erosion craters provides a wealth of new information about different aspects of crater formation like development of pores or extent of melting [42]. Figure 10c–f show a 3D-reconstruction of this particular crater obtained using the 3D reconstruction software Amira®. The whole oxide layer is disrupted during crater formation. Clearly, a deformation zone is visible underneath the crater. The extent of this deformation is at least on the order of the crater depth. The crater volume is on the order of 1,400 μm^3 (as calculated using the reconstructed volume, Fig. 10c). Figure 10d–f show coronal reconstruction views at 1.6 μm depth (coarse grained upper oxide layer), at 3.2 μm depth (fine-grained lower oxide layer) and at 6.5 μm (tip of the crater), respectively. In the last picture, the deformation around the tip can be qualitatively delineated.

For a better understanding of the nature of the deformation it would be essential to have the possibility for further characterization by EBSD, which would allow obtaining information about orientations of the crystalline lattice near the crater. Normally a FIB cross section through a structure element e.g. an inclusion is performed and imaged with the electron beam. To fulfill the optimal geometry condition for EBSD analysis (surface tilted to 70°) the polished surface has to be tilted and rotated to be imaged with the EBSD detector. However, in this case, an EBSD investigation of the surface is not possible because the remaining sample shields the detector from the surface. Therefore, EBSD investigations using FIB polishing are usually performed on surfaces confined to the edge of the sample.

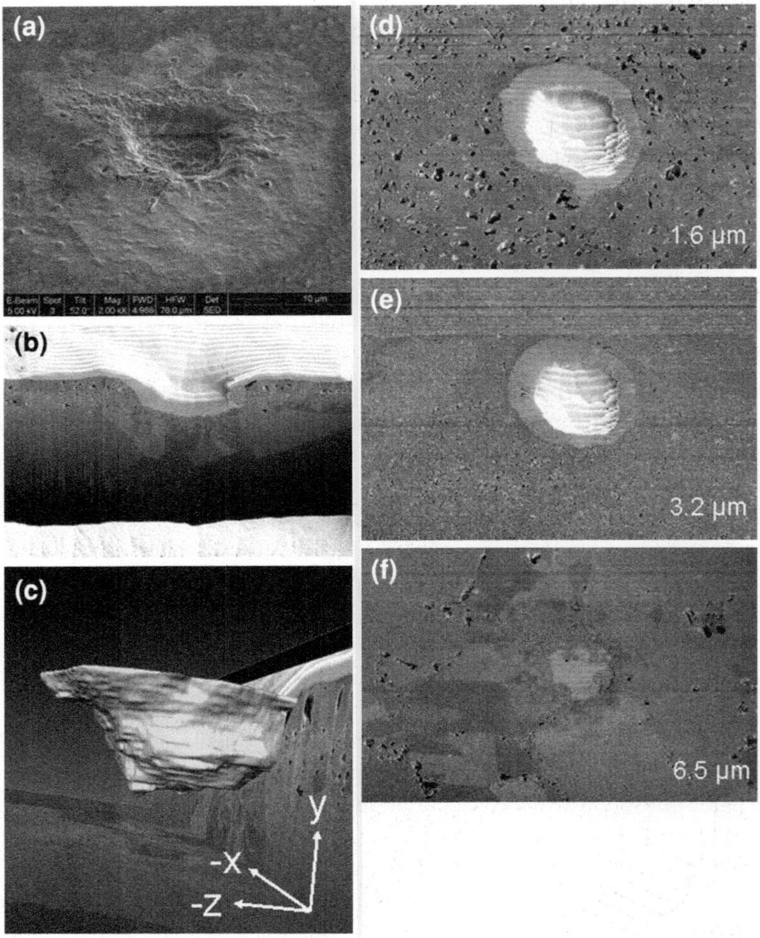

Fig. 10 Reconstruction of a plasma erosion crater using FIB tomography: In **a** the crater is shown before serial sectioning, **b** shows an example of one slice (cross section), in **c** the reconstructed crater together with the coordinate system (*x*, *y* are the coordinates of the SE images, *z* is the slicing direction) is depicted, and **d–f** show coronal reconstructions (plane *x–z*) at 1.6, 3.2 and 6.5 μm below the surface [15]

A mechanical pre-preparation of the sample to remove the material until reaching the crater, proved to be very time-consuming and sometimes results in sample loss (if polishing extends too far).

To avoid the geometry problem, one alternative is to cut the volume containing the feature to be investigated (volume of interest) by FIB milling and then positioning this volume element on the rim of the sample (Fig. 11). A Kleindiek MM3A micromanipulator was used for positioning the peace of material. Once positioned on the sample rim, the crater can be sectioned sequentially. A 45° pretilted holder was used and the geometry of EBSD polishing at the edge of the

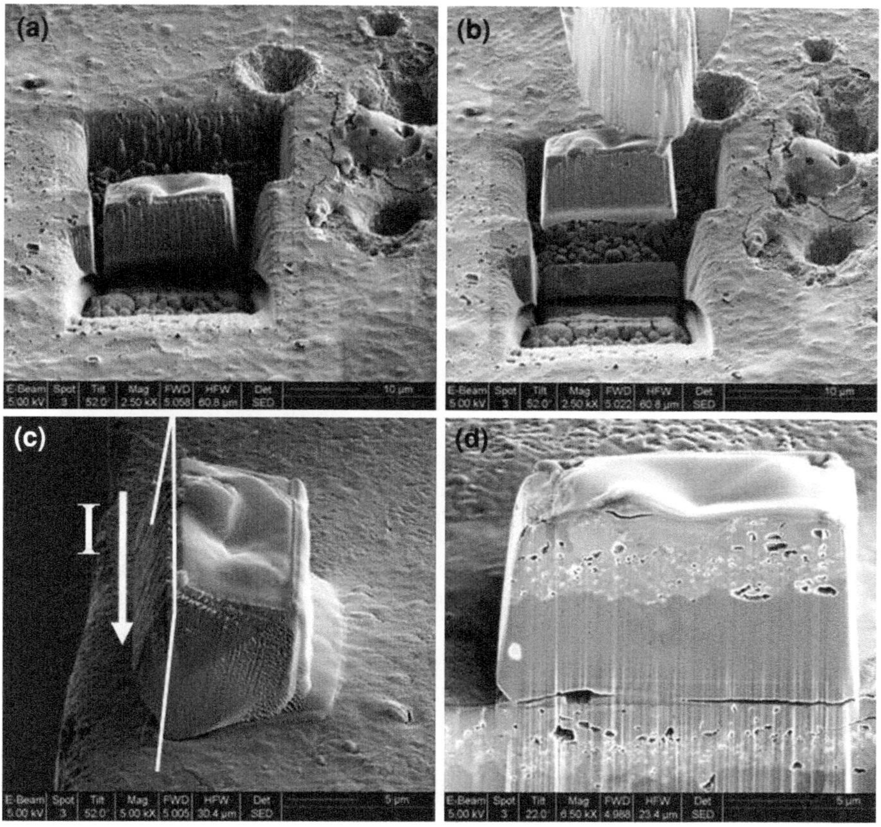

Fig. 11 Transfer of volume of interest for the EBSD tomography. The procedure is essentially the same as for TEM sample preparation. In **a** the volume of interest is cut free on all sides and from below and then lifted with the micromanipulator (**b**). In **c** the sample is positioned at the rim of the sample and polished with the ion beam in order to get an EBSD-quality surface (**d**) [15]

sample is similar to that as shown in [7]. In these systems the EBSD detector is mounted underneath the ion column, whereas in [43] a system with the EBSD detector on the opposite side was developed with the advantage that only a single tilt is necessary for a change between the FIB and the EBSD position.

Figure 12a shows a 3D reconstruction of the misorientation distribution using six slices in total, employing the Amira® 3D reconstruction software (one of the original cross section maps is presented in Fig. 12b). Shells corresponding to deformation between 1 and 2° (red) and between 2 and 5° (blue) together with the tip of the crater (yellow) and the NiO layer (transparent green) are reconstructed. Due to the large spacing of the individual slices (700 nm) the reconstruction has a relatively poor resolution in the z-direction. However, it becomes clear that the deformation (misorientation $\geq 1°$) extends to at least 6 μm below the tip of the crater and is therefore on the order of the thickness of the oxide layer and the depth of the crater, respectively. Areas of similar misorientation are emanating in a radial

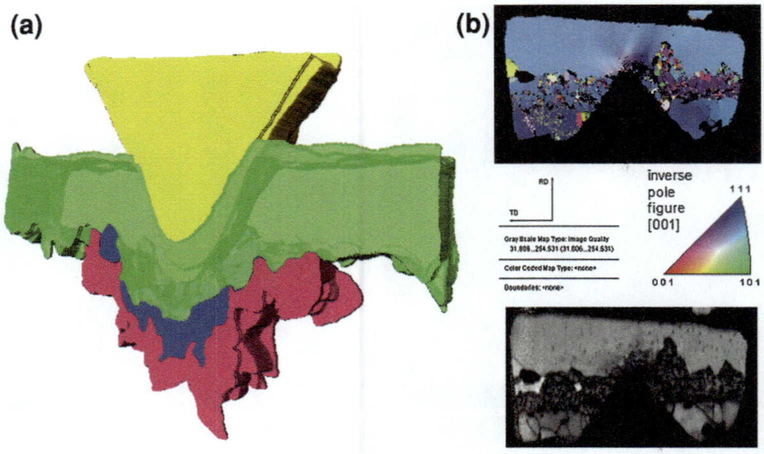

Fig. 12 **a** 3D reconstruction of part of the crater, oxide layer and the deformation zone. The deformation zone is shown in shells corresponding to misorientations between 1° and 2° (*red*) and between 2° and 5° (*blue*), the oxide layer is shown transparently (*green*); the upper oxide layer is not shown; **b** serial 3D IPF map obtained on serial sections through the plasma erosion crater (the image quality map below belongs to the IPF map shown on the *top*) the deformation zone is clearly visible in the slice

fashion from the tip of the crater, which seems to follow the direction of the heat flux during cooling by conduction into the sample. For an ideal point source at a semi-infinite surface the temperature distribution would follow a semi-spherical symmetry at each instant of time. Deformation following this symmetry of the temperature evolution would, in the ideal case, also exhibit this symmetry. The reason for the modulation of the deformation by a more radiant symmetry is assumed to be a superposition of different point sources at different depth and different times during the arc phase of the spark corresponding to the geometry of the crater tip. The orientation gradient itself is on the order of 3°/μm. Therefore no pronounced recrystallization is observed.

3.6 Other Applications

3.6.1 Porous-Nickel

Wicking structures are currently used in capillary pumped loop heat pipes for space applications. Typical wicking structures are fabricated with sintered metal powders (Ni, Ti and steel) and other low density porous structures like grooved surfaces, wire meshes and screens. The main purpose of a porous wick structure is to generate the capillary pumping pressure required to transport a working fluid along a two-phase heat transfer loop. The size of the inherent porosity and its quantification is an important task since the efficiency of the wick is directly

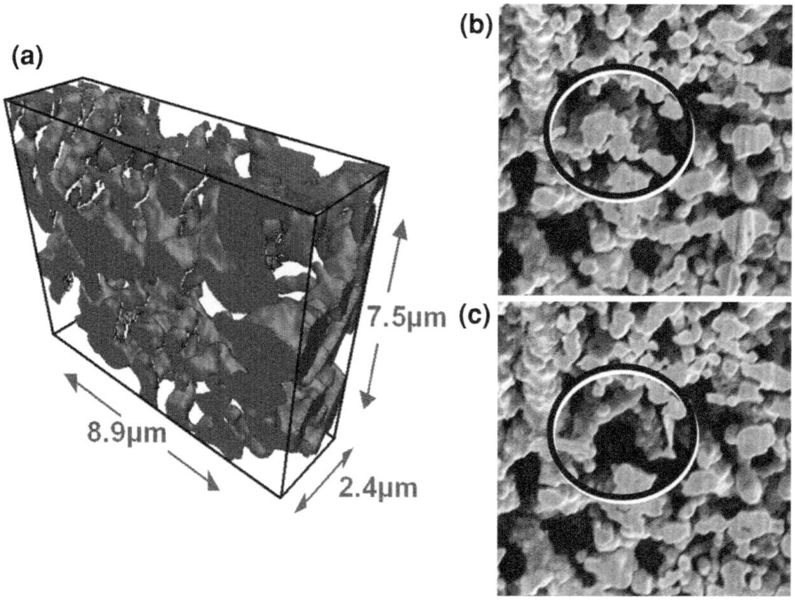

Fig. 13 **a** 3D FIB/SE reconstruction of a porous Ni structure; **b**, **c** detail of analyzed cross sections showing particle removal during the milling step. Voxel size (nm): $x = 29$, $y = 37$, $z = 60$ (slicing direction)

related to it. Typical values of 1–3 μm pore size and 50–60% porosity are desired. Furthermore, the modeling of the liquid flow through the wick requires detailed tridimensional information which can only be obtained by high detail characterization methods such as FIB tomography.

Figure 13a shows a FIB-tomograph of a porous Ni sample. In this case, SE images are able to reproduce the metal architecture with high voxel resolution. The analyzed volume presents an average porosity of 65 vol.% showing complete interconnection. The pore-canal like structure amounts to between 0.5 and 2 μm in diameter. The 3D reconstruction was obtained from 2D cross sections like in Fig. 13b, c. Bonding of the metallic particles during the sintering process appears randomly in any direction. Milling of such structures during preparation of the 2D cross section may affect sample integrity. As an example, a particle attached to main porous structure (Fig. 13b) is removed during the preparation of the next cross section (Fig. 13c). The bridge responsible for joining the metallic particle was removed at this step because of sample milling. This problem can be solved by the infiltration of the porous structure with a low-viscous resin or polymer at the liquid state, which after its solidification would act as supporter for the whole structure. Another important problem in the reconstruction of non-infiltrated porous materials is often the differentiation of milled surfaces from surfaces which are behind the plane of milling. This hinders any possibility of automatic segmentation and could lead to wrong segmentation. If infiltration is not possible, a careful manual segmentation seems to be the only solution for this problem.

3.6.2 Ordered SnO_2 Nanowires

The assembly of 3D arrangements of 1D nanostructures (nanowires, nanotubes, nanobelts and nanorods) has been shown to be an interesting approach for the design of functional structures. Arrays of SnO_2 nanowires (NW) for instance can be used for sensor or catalytical applications [44]. It is desired to represent in 3D the arrays of ordered NW. Since the NW have a diameter of 15–20 nm, only FIB tomography can be applied.

Tin oxide nanowires were synthesized by CVD of $[Sn(OtBu)_4]$ on Au coated $TiO_2(001)$ and $TiO_2(101)$ substrates. Contrary to the synthesis on other substrates (e.g. Al_2O_3), deposition on $TiO_2(001)$ and $TiO_2(101)$ substrates leads to an oriented growth with a mesh-like network of NW (see Fig. 14 [44]).

A tomograph of the NW arrays is shown in Fig. 15. The size of the region of interest is only $3.3 \times 4.6\ \mu m^2$, which allows for working with almost the maximal resolution possible: voxel size is $x = 3.3$ nm, $y = 4.4$ nm and $z = 10$ nm. The length and diameters of the NW could be manually measured with the software Amira®. For this sample, 33 NW were analyzed having a mean length of 309 nm with a standard deviation of 139 nm and a mean diameter of 30 nm with a standard deviation of 6 nm. Such results have been acquired for different samples and compared (to be published).

Although TEM tomography can offer quantitative characterization of nanoscopic 3D nanostructures, the 3D information is limited to less than 100 nm thin objects since the samples have to be thinned for electron transparency. The study

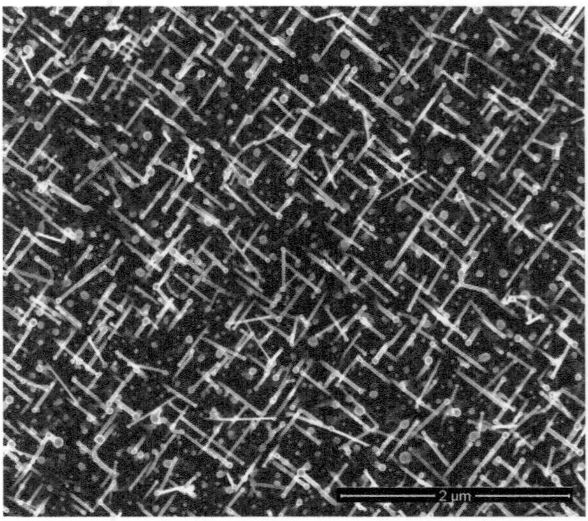

Fig. 14 SEM image of the SnO_2 nanowires arrays deposited on $TiO_2(101)$ substrate

Fig. 15 3D Reconstruction of the SnO_2 NW grown on a $TiO_2(101)$ substrate. The Au particles (light droplets) at the NW tip act as catalyst for the growth of the NW. Voxel size (nm): $x = 3.3$, $y = 4.4$, $z = 10$ (slicing direction). Size of the region described is $3.3 \times 4.6\ \mu m^2$

of ordered SnO_2 NW shows that FIB tomography can provide the 3D information of objects in large scale (micrometer size). It is possible to perform statistic analysis of nanowire growth direction and the distribution of catalysts.

Acknowledgements The authors would like to thank the following persons for the research collaborations in the different examples shown: M. Engstler, F. Miguel, S. Suarez (Saarland University), A. Lasagni (Fraunhofer IWS), H·P. Degischer (Vienna University of Technology), H. Shen, J. Pan, S. Mathur (University of Cologne), J. L. Garcia (Helmholz Zentrum Berlin), C. Holzapfel (Schleifring und Apparatebau GmbH). FAS and FM thank the EFRE Funds of the European Commission for support of activities within the AME-Lab project.

References

1. Spowart, J.: Automated serial sectioning for 3-D analysis of microstructures. Scr. Mater. **55**, 5–10 (2006)
2. Wucher, A., Cheng, J., Zheng, L., Winograd, N.: Three-dimensional depth profiling of molecular structures. Anal. Bioanal. Chem. **393**, 1835–1842 (2009)
3. Prewett, P.D., Mair, G.L.R.: Focused Ion Beams from Liquid Metal Ion Sources. Research Studies Press Ltd., Taunton (1991)
4. Orloff, J.: High-resolution focused ion beams. Rev. Sci. Instrum. **64**(5), 1105–1130 (1993)
5. Phaneuf, M.W.: Applications of focused ion beam microscopy to materials science specimens. Micron **30**, 277–288 (1999)
6. Sudraud, P., Ben Assayag, G., Bon, M.: Focused-ion-beam milling, scanning-electron microscopy, and focused-droplet deposition in a single microcircuit surgery tool. J. Vac. Sci. Technol. B Microelectron. Nanometer Struct. **6**(1), 234–238 (1988)
7. Young, R.J., Moore, V.M.: Dual-beam (FIB-SEM) systems. In: Gianuzzi, L.A., Stevie, F.A. (eds.) Introduction to Focused Ion Beams, Instrumentation, Theory, Techniques and Practice, pp. 247–268. Springer Science + Business Media, Inc., Boston (2005), Chapter 12
8. Steiger, W., Rudenauer, F., Gnaser, H., Pollinger, P., Studnicka, H.: New developments in spatially multidimensional ion microprobe analysis. Mikrochim. Acta **10**, 111–117 (1983)
9. Rüdenauer, F.G., Steiger, W.: A further step towards three-dimensional elemental analysis of solids. Mikrochim. Acta **76**, 375–389 (1981)

10. Dunn, D.N., Kubis, A.J., Hull, R.: Quantitative three-dimensional analysis using focused ion beam microscopy. In: Gianuzzi, L.A., Stevie, F.A. (eds.) Introduction to Focused Ion Beams, Instrumentation, Theory, Techniques and Practice, pp. 181–300. Springer Science + Business Media, Inc., Boston (2005)
11. Sakamoto, T., Cheng, Z., Takahashi, M., Owari, M., Nihei, Y.: Development of an ion and electron dual focused beam apparatus for three-dimensional microanalysis. Jpn J. Appl. Phys. **37**, 2051–2056 (1998)
12. Kato, M., Ito, T., Aoyama, Y., Sawa, K., Kaneko, T., Kawase, N., Jinnai, H.: Three-dimensional structural analysis of a block copolymer by scanning electron microscopy combined with a focused ion beam. J. Polym. Sci. B: Polym. Phys. **45**, 677–683 (2007)
13. Ray, S.S.: A new possibility for microstructural investigation of clay-based polymer nanocomposite by focused ion beam tomography. Polymer **51**, 3966–3970 (2010)
14. Matthijs de Winter, D.A., Schneijdenberg, C.T.W.M., Lebbink, M.N., Lich, B., Verkleij, A.J., Drury, M.R., Humbel, B.M.: Tomography of insulating biological and geological materials using focused ion beam (FIB) sectioning and low-kV BSE imaging. J. Microsc. **233**(3), 372–383 (2009)
15. Holzapfel, C., Soldera, F., Faundez, E., Muecklich, F.: Site-specific structural investigations of oxidized Ni samples modified by plasma erosion processes using the EBSD technique in combination with a FIB/SEM dual beam workstation. J. Microsc. **227**(1), 42–50 (2007)
16. Verhoeven, J.D.: Scanning electron microscopy. In: Whan, R.E. et al. (eds.) ASM Handbook, Volume 10 Materials Characterization, pp. 490–515. The Materials Information Society, Metals Park (1998), Fifth Printing
17. Lasagni, F., Lasagni, A., Holzapfel, C., Mücklich, F., Degischer, H.P.: Three dimensional characterization of unmodified and Sr-modified Al–Si eutectics by FIB and FIB EDX tomography. Adv. Eng. Mater. **8**(8), 719–723 (2006)
18. Schaffer, M., Wagner, J., Schaffer, B., Schmied, M., Mulders, H.: Automated three-dimensional X-ray analysis using a dual-beam FIB. Ultramicroscopy **107**(8), 587–597 (2007)
19. Schwartz, A.J., Kumar, M., Adams, B.L.: Electron Backscatter Diffraction in Materials Science. Kluwer Academic, New York (2000)
20. Zaefferer, S., Wright, S.I., Raabe, D.: Three-dimensional orientation microscopy in a focused ion beam-scanning electron microscope: a new dimension of microstructure characterization. Metall. Mater. Trans. A **39A**, 374–389 (2008)
21. Giannuzzi, L.A., Prenitzer, B.I., Kempshall, B.W.: Ion–solid interactions. In: Gianuzzi L.A., Stevie F.A. (eds.) Introduction to Focused Ion Beams, Instrumentation, Theory, Techniques and Practice, pp. 13–52. Springer Science + Business Media, Inc., Boston (2005), Chapter 2
22. Dunn, D.N., Hull, R.: Reconstruction of three-dimensional chemistry and geometry using focused ion beam microscopy. Appl. Phys. Lett. **75**, 3414–3416 (1999)
23. Inkson, B.J., Mulvihill, M., Möbus, G.: 3D determination of grain shape in FeAl-based nanocomposite by 3D FIB tomography. Scr. Mater. **45**, 753–758 (2001)
24. Stevie, F.A.: Focused ion beam secondary ion mass spectroscopy (FIB-SIMS). In: Gianuzzi L.A., Stevie F.A. (eds.) Introduction to Focused Ion Beams, Instrumentation, Theory, Techniques and Practice, pp. 269–280. Springer Science + Business Media, Inc., Boston (2005), Chapter 13
25. Smithells, C.J.: Equilibrium Diagrams, in Metals Reference Book, 6th edn. Butterworth & Co Ltd., London (1983), Ch. 11
26. Lasagni, F., Degischer, H.P., Papakyriacou, M.: Influence of solution treatment, Sr-modification and short fibre reinforcement on the eutectic morphology of Al–Si alloys. Prak. Met. **43**(10), 505–519 (2006)
27. Lasagni, F., Lasagni, A., Marks, E., Holzapfel, C., Mücklich, F., Degischer, H.P.: Three-dimensional characterization of 'as-cast' and solution-treated $AlSi_{12}$(Sr) alloys by high-resolution FIB tomography. Acta Mater. **55**(11), 3875–3882 (2007)
28. Nafisi, S., Ghomashchi, R.: Effects of modification during conventional and semi-solid metal processing of A356 Al–Si alloy. Mater. Sci. Eng. A **415**(1/2), 273–285 (2006)

29. Dons, A.L., Pedersen, K.O., Voje, J., Mæland, J.S.: Heat treatment and impact toughness properties of AlMgSi foundry alloys. Aluminium **81**(1/2), 98–102 (2005)
30. Mondolfo, L.F.: Al Alloys: Structures and Properties. Butterworth, London (1976)
31. Degischer, H.P., Knoblich, H., Knoblich, J., Maire, E., Slavo, L., Suery, M.: Proceedings of the 12 International Metallographie Tagung, Leoben, Austria, Prak. Metallographie Sonderband **38**, 67–74 (2006)
32. Qin, Q.D., Zhao, Y.G., Cong, P.J., Zhou, W., Xu, B.: Semisolid microstructure of Mg_2Si/Al composite by cooling slope cast and its evolution during partial remelting process. Mater. Sci. Eng. A **444**, 99–103 (2007)
33. Lasagni, F., Lasagni, A., Engstler, M., Degischer, H.P., Mücklich, F.: Nano-characterization of cast structures by FIB-tomography. Adv. Eng. Mater. **10**(1/2), 62–66 (2008)
34. Amira 3.1: User's Guide and Ref. Manual, Konrad-Zuse-Zentrum für Informationstechnik Berlin (ZIB), Germany
35. Lasagni, F., Lasagni, A., Holzapfel, C., Engstler, M., Mücklich, F.: 3D microstructural study of $AlSi_8Mg_5$ alloy by FIB-tomography. Prak. Met. **9**(10), 487–499 (2010)
36. Ruppi, S., Halvarsson, M.: TEM investigation of wear mechanisms during metal machining. Thin Solid Films **353**, 182–188 (1999)
37. Pitonak, R., Garcia, J., Weissenbacher, R., Udier, K.: Austrian Patent AT503050 B1 (2007)
38. Garcia, J., Pitonak, R., Weißenbacher, R., Köpf, A., Soldera, F., Suarez, S., Miguel, F., Pinto, H., Kostka, A., Mücklich, F.: Design and characterization of novel wear resistant multilayer CVD coatings with improved adhesion between Al_2O_3 and Ti(C, N). Adv. Eng. Mater. **12**, 929–934 (2010)
39. Garcia, J., Pitonak, R., Weissenbacher, R., Köpf, A.: Production and characterization of wear resistant Ti(C,N) coatings manufactured by modified chemical vapor deposition process. Surface & Coatings Technology **205**, 2322–2327 (2010)
40. Kak, A.C., Slaney, M.: Principles of Computerized Tomographic Imaging. IEEE Press, New York (1988)
41. Soldera, F., Mücklich, F., Kaiser, T., Hrastnik, K.: Description of the discharge process in spark plugs and its correlation with the electrode erosion patterns. IEEE Trans. Veh. Technol. **53**, 1257–1265 (2004)
42. Jeanvoine, N., Holzapfel, C., Soldera, F., Mücklich, F.: 3D investigations of plasma erosion craters using FIB/SEM dual-beam techniques. Pract. Metallogr. **43**(9), 470–482 (2006)
43. Konrad, J., Zaefferer, S., Raabe, D.: Investigation of orientation gradients around a hard Laves particle in a warm-rolled Fe_3Al-based alloy using a 3D EBSD-FIB technique. Acta Mater. **54**(5), 1369–1380 (2006)
44. Pan, J., Shen, H., Werner, U., Prades, J.D., Hernandez-Ramirez, F., Soldera, F., Mücklich, F., Mathur, S.: Heteroepitaxy of SnO_2 Nanowire Arrays on TiO_2 Single Crystals: Growth Patterns and Tomographic Studies. Submitted to Chemistry of Materials (2011)

Atom Probe Tomography: 3D Imaging at the Atomic Level

D. Blavette, F. Vurpillot, B. Deconihout and A. Menand

Abstract Atom probe tomography (APT) is the only approach able to map out the 3D distribution of chemical species in a material at the atomic-scale. The instrument provides quantitative measurements of local chemical composition in a small selected volume at the nm scale. The in-depth spatial resolution reaches a few tens of picometres enabling atomic layers to be imaged. The lateral resolution is however limited to a fraction of a nanometre, precluding therefore the crystal lattice to be fully reconstructed in the general case. APT has been applied to number of issues in physical metallurgy including precipitation, segregation of impurities to lattice defects, ordering, magnetic multilayers. Up to the beginning of this century, APT has been mainly limited to metals. The implementation of ultrafast laser pulses instead of high-voltage pulses to field evaporate surface atoms has now opened APT to the analysis of semi-conductors or oxides that are key materials in microelectronics. With the introduction of ion milling using Focussed Ion Beam (FIB), the instrument has now gained a key place in nanosciences. Salient results were hence recently produced on tunnel junctions, nanowires, oxides, dopant distribution in semiconductors and nano-transistors. The potential of APT as well as both limitations and future prospects will be discussed on the basis of some selected illustrations.

D. Blavette (✉), F. Vurpillot, B. Deconihout and A. Menand
University of Rouen, Groupe de Physique des Matériaux, Institut CARNOT ESP, UMR CNRS 6634, BP 12, 76801, St Etienne du Rouvray Cedex, France
e-mail: didier.blavette@univ-rouen.fr

D. Blavette
Institut Universitaire de France, Versailles, France

F. A. Lasagni and A. F. Lasagni (eds.), *Fabrication and Characterization in the Micro-Nano Range*, Advanced Structured Materials, 10,
DOI: 10.1007/978-3-642-17782-8_9,

1 Introduction

Atom probe tomography (APT) is the only approach able to map out the 3D distribution of chemical species in a material at the atomic-scale. The instrument [so-called 3D atom-probe (3DAP)] is an extension in 3D of the atom-probe field ion microscope (APFIM), an instrument designed in the late 1960s by Müller et al. [1]. The principle is based on the field evaporation of surface atoms of the specimen and the chemical identification of field evaporated ions by time-of-flight mass spectrometry. The position of atoms at the sample surface is derived from the impact position of ions striking the detector. The two first prototypes were designed successively at the Universities of Oxford and Rouen, the French version being the Tomographic atom probe (TAP) [2, 3]. This latter is commercialized by CAMECA. Another commercial instrument was designed latter by the American society IMAGO (LEAP, local electrode atom probe). This latter company has now joined CAMECA. The in-depth resolution, independent of technologic details of these instruments, reaches 10 pm [4, 5]. Unfortunately, the lateral resolution (at the sample surface) is much worse and is a few tenths of a nanometre in the best cases. Lattice reconstruction was achieved only in very limited cases such as some pure metals [6].

A major advantage of APT is its quantitativity. The local composition in a small selected region of the analyzed volume is simply derived from the number of atoms of each observed species. This is a big advantage compared to number of other instruments. No calibration is required. Thanks to its high depth resolution, atomic planes can be imaged and the chemical order can be exhibited in ordered regions. Among the more salient results obtained, let us mention the imaging of Cottrell atmospheres (tiny clouds of impurity atoms around dislocations in crystals). The concept of “atmospheres” was introduced by Cottrell and Bilby [7] in 1949 to explain the role of impurities in the plastic. Thanks to 3DAP, a Cottrell atmosphere in FeAl intermetallics was imaged in 3D and at the atomic scale for the first time [8]. More generally, APT was shown to be a very powerful instrument in physical metallurgy, particularly for the investigation of the early stages of decomposition of alloys.

3DAP was up to recently restricted to good conductors (metallic alloys or low-resistance oxides). To overcome this limitation, the laboratory designed a new generation of 3DAP in which the material removal is ensured by ultrafast laser pulses (< 500 fs). This new instrument, namely the laser assisted wide-angle tomographic atom probe (LAWATAP, CAMECA) has opened the field of application of the technique to materials with a poor electrical conductivity such as semi-conductors and oxides, which are key materials in micro-electronics [9]. In this instrument, the evaporation of surface atoms is assisted by femtosecond pulses instead of voltage pulses [10]. In addition, a wider field of view is achieved so that larger area of analysis are obtained ($\sim 100 \times 100$ nm^2) improving therefore statistics together with shortening the number of analyses required to get the relevant information. A similar instrument using 10 ps laser pulses was developed by IMAGO.

With the impressive progresses made in the miniaturization of integrated circuits, microelectronics occupies a particular place in Nanoscience. The size of last generation nano-transistors is a hundred nanometers. The SIMS is the reference tool in microelectronics for dopant profiling in semiconductors. However, for nano-transistors SIMS faces to its ultimate limits and 3DAP is appealed to play an important role [11]. The Moore law already faces new physical limits and new challenges for which APT could bring new and relevant results. More generally, APT appears to be a very powerful approach in nanoscience in particular for the investigation of nanowires and multilayers including tunnel junctions containing highly resistive oxide layers [12].

2 Basic Principles

The principle of the tomographic atom probe is based on the pulsed field evaporation of surface atoms and the identification of field evaporated ions by time of flight mass spectrometry. The position of atoms at the sample surface is derived from the impact coordinates onto the detector. A simple point projection is involved (Fig. 1). The in-depth investigation of the sample is provided by the layer-by-layer evaporation of the specimen.

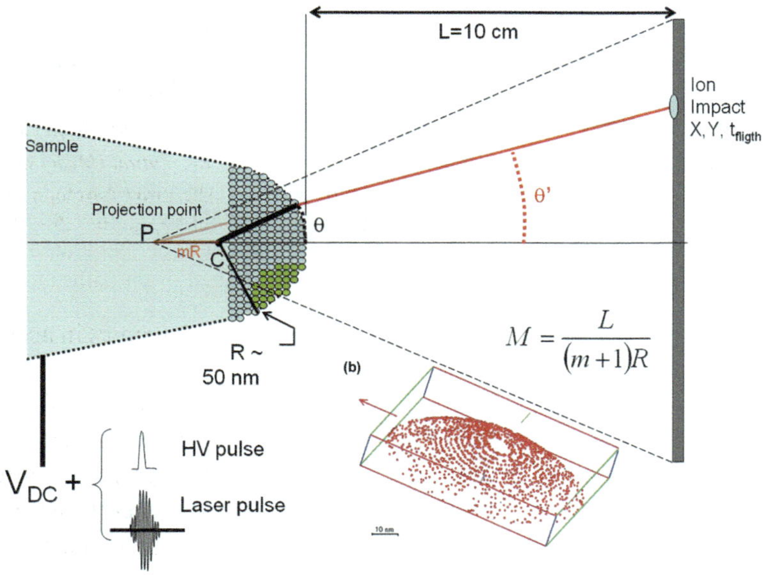

Fig. 1 The principle of the tomographic atom probe. In the laser assisted tomographic atom probe (LaWaTAP), electric pulses are replaced by ultrafast laser pulses (pulse width < 500 fs)

The high electric field (F) required (a few tens V/nm) to field evaporate surface atoms is obtained by applying a high electric voltage (V_0) to the specimen prepared in the form of a sharply pointed needle [the tip radius (R) is close to 50 nm, $F \sim V_0/R$]. Whereas electrochemical techniques were formerly used to prepare tips from bulk materials, ion milling (focused ion beam, FIB) is now a common approach to prepare sharply pointed needles from a large variety of samples including multilayers, nanotransistors, nanopowders, buried interfaces [13]. Clearly, the introduction of FIB techniques has opened the instrument to nanosciences.

As shown in Fig. 1, the specimen has the shape of a truncated cone terminated by a spherical cap. This means that the tip radius slightly increases as surface atoms are removed. The electric field F created at the tip surface is simply given by $F = V_0/\beta R$ where β is the field factor ($\beta = 1$ for an ideal sphere). The field factor β (2–8) depends on the shank angle and on the electrostatic environment of the specimen. Because of the shank angle, a slow increase of the applied voltage is generally necessary in order to keep constant the applied field and to compensate for the increase of the tip radius. V is automatically increased during analyses in such a way to maintain the detection rate constant (typically 0.01 ion/pulse).

The evaporation flux is thermally activated and is given by $\Phi = \Phi_0 \exp(-W(F)/kT)$. The activation energy $W(F)$ decreases almost linearly with the applied field (F). Increasing the tip temperature or the applied field therefore promotes field evaporation. It is worth mentioning that analyses are always conducted at cryogenic temperatures. Specimens are cooled down to temperatures ranging from 20 to 100 K in order to get quantitatitive composition data and to prevent diffusion of surface atoms in order to preserve the spatial resolution. At cryogenic temperatures high electric fields of 10–100 V/nm are required to field evaporate surface atoms.

The chemical identity of atoms is derived from the time-of-flight of chemical species that are field evaporated in the form of multiple charged ions. High voltage pulses V_p superimposed to V_0, lead to the field evaporation of surface atoms as ions that are repelled from the tip surface. Depending on the design of the instrument the repetition rate can vary from 1 to 100 kHz. Field evaporated ions are detected on a single particle sensitive detector (channel plates). The generated signals provide the time of flight (t) of chemical species. For typical flight paths ($L \sim$ 0.1–0.5 m), t is of the order of a few hundred of nanoseconds. The mass to charge ratio of ions (m/n) is derived from the simple equation $1/2\, mv^2 = n\, e(V_0 + V_p)$ with $v = L/t$ and n the charge state. The detection efficiency of channel plates (electron multiplier arrays—10^7 electrons/ion impact) used in the detector is close to the open area of the device ($\sim$ 60%). Last improvements brought about to the detector rise the efficiency to 70%.

Under proper conditions (tip temperature below 80 K, high pulse fraction ($V_P/V_0 = 20\%$), the ionization rate is the same for every chemical species. This ensures the quantitativity of concentration measurements. It is worth mentioning that using smaller pulse fraction may lead to the preferential evaporation of low evaporation field atoms (low binding energy atoms) at the DC voltage. Let us consider an AB alloy where B atoms have a lower evaporation field. If B atoms are preferentially evaporated at the DC voltage, then they will not be detected in coincidence with pulses. Then the amount of B atoms will be underestimated. Consequently, the measured atomic fraction of B atoms will be smaller than expected.

The instrument if always combined with a field ion microscope (FIM). FIM is based on the field ionization of a rare gas near the tip surface [14]. Ion beams originating from protruding surface atoms (high field) received on a screen makes it possible to get a magnified image of the specimen surface resolved on the atomic scale. FIM experiments combined with atom-probe proved that the ionization efficiency was close to 100%. Surface atoms were observed using FIM in proportion as they were removed from the surface by field evaporation. The fraction of removed atoms in the observed area that were detected was found very close to the detection efficiency.

The use of a time-resolved position sensitive detector (PSD) makes it possible to position ion impacts and to calculate the position from which atoms originate at the tip surface (Fig. 1). Several types of detectors were implemented on the atom-probe (multianode detector, the first French generation of tomographic atom probe [15], time resolved CCD detector [16]). Delay line based detectors are now most often used [17]. PSDs generally fail to assign correct impact positions for multiple events in particular for ions which both have the same m/n and are close to each other on the detector (at a delay < 1.5 ns and at a distance < 1.5 mm with recent detectors [17]). Small detection rates are generally used (0.01 ion/pulse) in order to reduce multiple events that may bias both positions and composition data.

A simple point projection, close to a stereographic projection, is involved in the reconstruction of the analyzed volume (Fig. 1). The magnification (M), close to 10^7, is given by: $M = L/(m + 1)R$, where m defines the position of the centre of projection ($CP = mR$ with $m \sim 0.6$). The position of atoms at the tip surface (X) is roughly derived from that of impacts (X_D) using the equation $X = X_D/M$. More details may be found elsewhere [18].

The actual reconstruction procedure is obviously slightly more complex. The impact angle θ' (Fig. 1) is proportional to the emission angle θ of the atom at the tip surface [$\theta' \sim \theta/(m + 1)$]. Due to the symmetry of the tip, the azimuth angle $\varphi' = \varphi$. Assuming a radius of curvature R, the Cartesian coordinates of the atom is given by the classical set of equations:

$$\begin{aligned} x &= R\sin(\theta)\sin(\varphi) \\ y &= R\sin(\theta)\cos(\varphi) \\ z &= R(1 - \cos(\varphi)) \end{aligned} \tag{1}$$

The radius of curvature is deduced from the applied voltage (V): $R = V/F\beta$. As already said, because of the shank angle, the tip radius (R) increases with the analyzed depth (z) so that to ensure a constant evaporation field. This results in a decrease of the magnification (M) in proportion as the tip is field evaporated.

Thus the analyzed area ($A = A_D/M^2$) is an increasing function of R and consequently of the applied voltage V ($V = F\beta R$):

$$A(V) = A_D\left(\frac{(m+1)V}{F\beta L}\right)^2 \tag{2}$$

The depth scale (z) is generated simply derived from the expression of the analyzed volume related to the detection of δn ions:

$$\delta n\Omega = QA(V)\delta z \tag{3}$$

δz is the elemental depth analyzed related to the detection of δn ions. This latter expression is only valid for a small value of δz (i.e. small δn) for which $A(V)$ can be considered as constant. z is finally calculated by summing elemental depth δz.

Last generations of instruments have a larger field of view as compared to the previous designed in the early 1990s (40–100 nm). This could be achieved through a reduction of the flight path (L), and the implementation of ion optics and PSD with improved performance.

The maximum depth that can be explored with 3DAP is a few hundreds nm. It is in theory limited by the maximum voltage (V) that can be applied to the tip. A small shank angle makes it possible in principle to analyze a larger depth. However the main limiting factor is due to tip fracture: the high electrostatic pressure that is exerted often leads to fracture and this generally occurs before attaining the maximum voltage (20 kV) that can be applied to the sample.

The volumes that can be analyzed by APT ($50 \times 50 \times 100$ nm^3) typically contain a few tens millions of atoms. An example is given in Fig. 2. This 3D reconstruction exhibits the presence of small Al-enriched γ' precipitates (7 nm) finely dispersed in a Cr enriched matrix in a model nickel base superalloy. These small precipitates were found to contain around 18 at.% of Al and a few at.% of Cr whereas the parent γ phase contained around 20 at.% of Cr and a few at.% of Al.

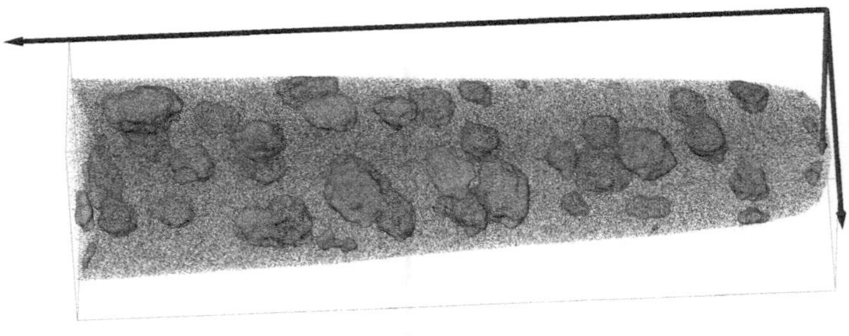

Fig. 2 3D analysis of a model nickel base superalloy using atom probe tomography. Only Al and Cr atoms are represented. Nickel is omitted for clarity. Small Al-enriched precipitates (7 nm in diameter) embedded in a Cr-enriched parent phase are evidenced. The transparent envelop represents the iso-surface of Al concentration with a 10% threshold. The volume consists of about 10^7 atoms ($40 \times 40 \times 150$ nm^3)

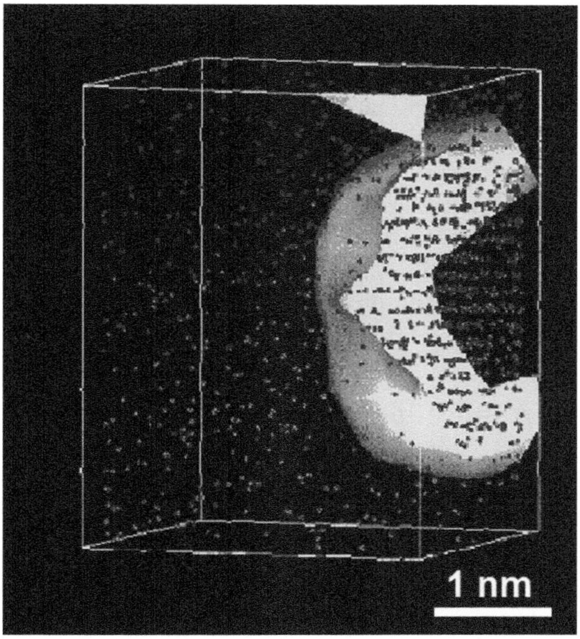

Fig. 3 Zoom of a small γ' precipitate observed in Fig. 2 showing its $L1_2$ ordered structure (Ni_3Al). Only Al is represented. Al-enriched (001) planes alternate with Al-depleted planes. The (001) planes interspacing is close to 0.36 nm (the lattice parameter). The diameter of the precipitate is equivalent to 20 atomic planes ($\sim$ 7 nm)

One of the main advantages of APT is its high spatial resolution. The depth resolution (along the tip axis), mainly controlled by the screening distance of the electric field, is close to 0.1 nm. This makes it possible to image low index planes. As shown in Fig. 3, the (001) planes of γ' precipitates are clearly imaged. Al-enriched planes alternate with Al-depleted planes depicting the ordered nature of $L1_2$ Ni_3Al precipitates. The averaged composition of each type of plane can be derived from such an image so that the order parameter and the preferential sites of additional elements can be estimated. Care should be taken to preferential retention of high evaporation field surface atoms (e.g. refractory elements) that can be retained from one plane to the following. This retention can be observed using FIM. This artifact will result to biased site occupancy frequencies.

Whereas the depth resolution of APT is excellent, the lateral resolution (parallel to the sample surface) is far from being as good. In one phase materials, it is close to 0.3 nm, making therefore impossible the reconstruction of the crystal lattice in 3D. It is the physics of field evaporation and not the performance of the position sensitive detector that limits the lateral resolution. Aberrations in the ion trajectories very close to the tip surface indeed deteriorate the resolution. Atomic simulations of ion trajectories that were conducted clearly indicate that it is the subtle movements of ions leaving the surface that are responsible for the dispersion of impact positions on the detector [19, 20]. These aberrations depend on the small range neighborhood of the atom leaving the surface and on the local atomic structure and arrangement, as much information that is not a priori known [21, 22]. Several approaches were developed to deal with this issue [23–25].

Numerous methods were developed to interpret APT data including 3D autocorrelation functions, cluster identification and erosion procedures based on mean separation methods or concentration criteria, contingency methods, proximity histograms, spatial distribution maps [4, 6, 26–32]. Fourier transform can be used to filter images in the reciprocal space, to make dark field images of ordered precipitates [33, 34]. Bright field images can also be obtained in order to get spatial information related to small precipitates. Distribution of nearest neighbor distances is an elegant way to get the composition of small clusters that has been recently developed [35, 36]. Pair correlation functions are also of great interest to get not only the size of precipitates but also their composition [37, 38].

3 Applications in Physical Metallurgy: Potential and Limits

Because of its high spatial resolution APT is particularly suitable for the measurement of the composition in regions as small as 1 nm (small precipitates [39], grain-boundary (GB) segregation, Segregation to APBs or staking faults (Suzuki effect) [40], concentration gradients close to interfaces, solute enrichment to line defects…). Such information is essential for alloys of industrial interest. For instance the phase composition is known to control the lattice misfit between γ and γ' phases in superalloys which in turn is a key parameter for creep performance of superalloys implemented in aircraft turbines. The nominal composition of a polycrystalline superalloy (N18) developed for turbine disks as well as the phase composition as derived from APT analyses are provided in Tables 1 and 2 [41]. As shown, γ' precipitates are Al, Ti, and Hf enriched compared to the γ solid solution that contains a large amount of Cr, Co and in less extent Mo. The volume fraction of γ' precipitates as derived from the phase composition using the lever rule ($f = (C_n - C_\gamma)/(C_{\gamma'} - C_\gamma)$) was found close to 57%.

From both a fundamental point of view and for the control of microstructure genesis, the early stages of precipitation and the kinetics pathway are key

Table 1 Nominal composition of the superalloy (N18)

Elément	Ni	Cr	Al	Ti	Mo	Co	Hf	B	C	Zr	Fe
Atomic %	54.42	12.3	9.15	5.11	3.77	14.82	0.16	0.083	0.075	0.018	0.11
Weight %	57.05	11.4	4.41	4.37	6.47	15.6	0.52	0.016	0.016	0.03	0.11

Table 2 Composition of both γ and γ' phase in atomic % in superalloy N18

At.% $\pm 2\sigma$	Ni	Cr	Al	Ti	Mo	Co	Hf	B	C	Fe
γ	38.28	25.71	1.85	0.25	8.36	25.28	0.01	0.04	0.01	0.20
18,400 Ions	0.72	0.64	0.20	0.07	0.41	0.64	0.02	0.03	0.02	0.07
γ'	67.71	1.60	12.95	8.65	2.20	6.64	0.21	0.01	0.00	0.03
84,784 Ions	0.32	0.09	0.23	0.19	0.10	0.17	0.03	0.01	0.00	0.01

Statistical fluctuations caused by sampling errors are given by the standard deviation σ ($\Delta C = 2\sigma$ with $\sigma = \sqrt{(C(1 - C)/N)}$ and N is the number of ions)

questions. Impressive results on model superalloys were for instance obtained by Schmuck and more recently by Seidman [42, 43]. Observations were confronted to nucleation-growth-coarsening theories. It is important to note that volumes reconstructed by APT ($50 \times 50 \times 100$ nm^3) have comparable dimensions than those simulated by kinetic Monte Carlo simulations (rigid lattice model with vacancy exchange) and microstructure features are displayed at the same scale, the atomic scale. It was therefore quickly realised that a combined investigation could be very beneficial in particular to calibrate kinetics and thermodynamic parameters of simulations and to validate simulation predictions [44]. Such fundamental issues are of utmost importance for the long term evolution of commercial steels. APT has played a crucial role in the investigation of the spinodal decomposition of the ferritic phase (Fe–Cr) of duplex stainless steels implemented in primary coolant pipes of PWR nuclear plants [45]. The segregation of solutes like boron to GBs is another crucial issue as this element is added to the chemistry of polycrystalline superalloys (disks of engines) in order to prevent intergranular fracture. A review of the many original information brought about to nickel base superalloys may be found in a recent compilation article [46].

One force of APT is that from reconstructed images such as that shown in Fig. 2, it is possible to get the local composition in a small selected region with arbitrary dimensions. The choice of a too small sampling volume (under 1 nm^3) will evidently lead to considerable statistical fluctuations. The amplitude of sampling errors ($\delta C = 2\sigma$) is given by the standard deviation $\sigma = (C(1 - C)/N)^{1/2}$ with $N = QV_m/\Omega$ with Q the detection efficiency (~ 0.6), V_m the volume over which the measurement is achieved and Ω the average atomic volume (~ 0.012 nm^3 in Ni). Fluctuations increase when approaching the equiatomic composition ($C = 0.5$). For $V_m = 1$ nm^3, the number of detected ions is only $N \sim 40$ atoms. For $C = 0.5$, fluctuations reach $\delta C = 2\sigma \sim 0.2$. More details on statistics may be found elsewhere [47, 48].

The assessment of the composition of very small precipitates poses not only the problem of statistical fluctuations but also that of the spatial resolution. Experiments as well as simulations have demonstrated that ion trajectory aberrations are much more pronounced in two-phase alloys compared to one phase materials. This problem arises when the two phases have different evaporation fields (F). For a given applied voltage (V), the local radius $R = V/F\beta$ will be smaller when F is larger. High field precipitates (resp. small) will therefore develop during their evaporation a smaller radius of curvature (resp. larger) compared to the surrounding parent phase [49]. These local magnification effects can lead to aberrations as high as 1 nm close to interfaces. This makes it very difficult the quantitative assessment of the composition of precipitates particularly when their sizes approach 1 nm. Trajectory overlaps close to interfaces lead to a mixed region and apparent broad interfaces. For a low field precipitate (focusing effect), solvent ions coming from the surrounding matrix reach the precipitate region. The solvent content is thus overestimated (lower solute content). Correction procedures accounting for these local magnification effects have been developed [50]. Combined investigation of ultra fine precipitation in

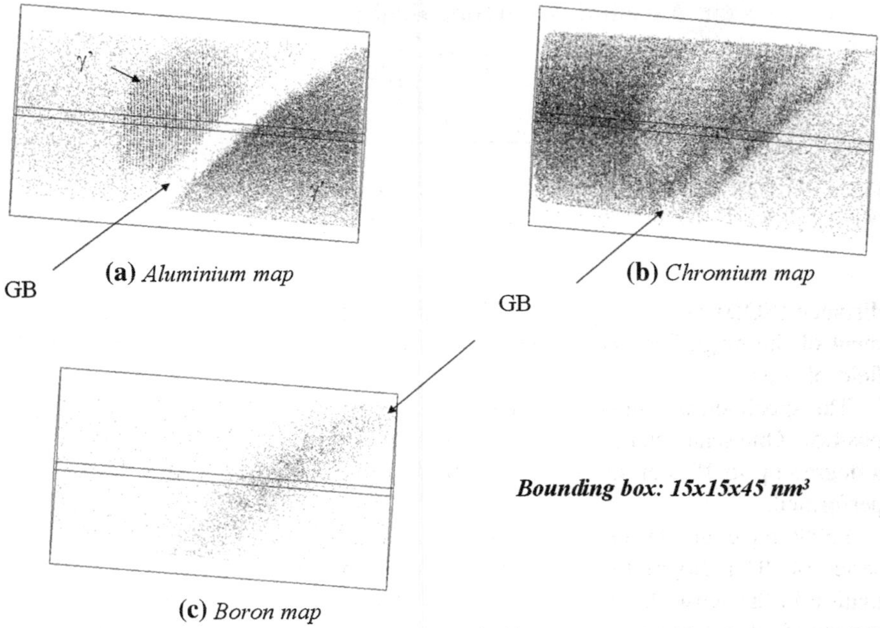

Fig. 4 Elemental maps showing the segregation of boron to the grain boundary in N18 superalloys. **a** only Al atoms, **b** Cr atoms, **c** boron atoms. These images show the presence of a small Ni_3Al γ' precipitate on the left side of the boundary. The γ' phase is also present on the right side (the size of the particle is larger that the size of the volume). Note the strong depletion of Al in the grain boundary and the enrichment of Cr typical of the γ solid solution (courtesy of Cadel [54])

Al based alloys using both APT and small angle X ray diffusion was achieved by A. Deschamps et al. [51].

Similar problems arise when investigating GB segregation [52]. Boron enriched GBs in N18 superalloys appear broader than expected for equilibrium segregation. An example is provided in Fig. 4 where the boron enriched layer spread over 5 nm. Experiments demonstrate that the width of the boron enriched zone tends to its theoretical value (0.5 nm) when the GB is perpendicular to the tip axis [23, 53]. In this situation, it is not any more the lateral resolution that controls the reconstructed image of the segregated GB but the depth resolution that is better than 0.1 nm. One consequence of the occurrence of such local magnification effects is that because of the broadening of enriched layer, the local concentration of boron is found smaller than expected. Composition can be corrected assuming that boron atoms imaged in the GB region were in reality concentrated in a thin layer of given thickness (0.5 nm for equilibrium segregation). In contrast to the local composition, the Gibsian interfacial excess expressed as the excess number of impurities segregated per unit of area of GB is not biased. This latter quantity has an important role in the thermodynamics of interfacial segregation [54].

4 The Laser Assisted 3D Atom Probe

The last few years have been the witness of a major breakthrough. Formerly limited to metals or good conductors, the implementation of ultra-fast pulsed laser to the instrument opened APT to semi-conductors or oxides and as a consequence to the important domain of microelectronics and nanosciences (e.g. tunnel junctions…). In the first generations of voltage pulsed 3D atom-probes, materials with low electrical conductivity could not be analyzed properly. High voltage pulses could not be transmitted properly to the tip apex. The broadening of the HV pulse transmitted to the tip apex as well as the lowering of its amplitude lead to very deteriorated mass spectra that were often impossible to index properly. For highly resistive materials, it was even impossible to get any data. In the last generation of tomographic atom probe (LaWaTAP, laser assisted wide angle tomographic atom probe—CAMECA), HV pulses have been replaced by femtosecond laser pulses. These pulses gives rise to a very rapid thermal pulse that promotes the field evaporation of surface atoms. This made it possible to analyze bad conductors like semiconductors or oxides that are key materials in microelectronics.

25 years ago, Kellogg and Tsong were the first to implement a pulsed laser on a one-dimensional atom-probe [55]. However, ultrafast pulsed laser was never implemented to 3D atom-probe until recently. We showed the first results in 2004 at the International Field Emission Symposium [56]. We demonstrated that the use of ultrafast laser pulses allows to field evaporate surface atoms in intrinsic silicon. As expected, the very short duration of the laser pulse significantly weakens the ion energy distribution leading to a high mass resolution without the need of time focusing or energy compensating devices. The duration of light pulses is a few hundreds fs and their energy is in the range 0.1–1 μJ according to the focalization of the beam on the tip.

The mass spectrum of a Si–SiO_2–Si sandwich is provided in Fig. 5. Silicon is mainly observed as doubly charged ions. A proportion of Si^+ is however detected. The charge state of ions depends on the electric field. Surface atoms are originally field evaporated as singly charged ions. When the electric field is sufficient, field evaporated ions can undergo an post-ionization by tunneling close to the surface [57]. When increasing the applied voltage (V_0), doubly or triply charged ions are detected. When a smaller DC voltage is needed in order to prevent preferential evaporation, the energy of laser pulses need to be increased in order to maintain the evaporation rate. This results in a larger proportion of singly charged ions in proportion as the DC voltage is decreased.

Successful analyses of highly resistive materials were performed such as intrinsic silicon, or even silicon oxide (Fig. 5). Mass spectra demonstrate that the use of laser pulses greatly improves both the mass resolution ($M/\delta M \sim 700$ (FWHM) with a flight length of 10 cm) and the signal-to-noise ratio. Using electrostatic lens, both the flight times of ions as well as the analyzed area are increased. This results in larger analyzed volumes and better statistics as well as a drastically improved mass resolution ($M/\Delta M \sim 3{,}000$ (FWHM)). Increasing the

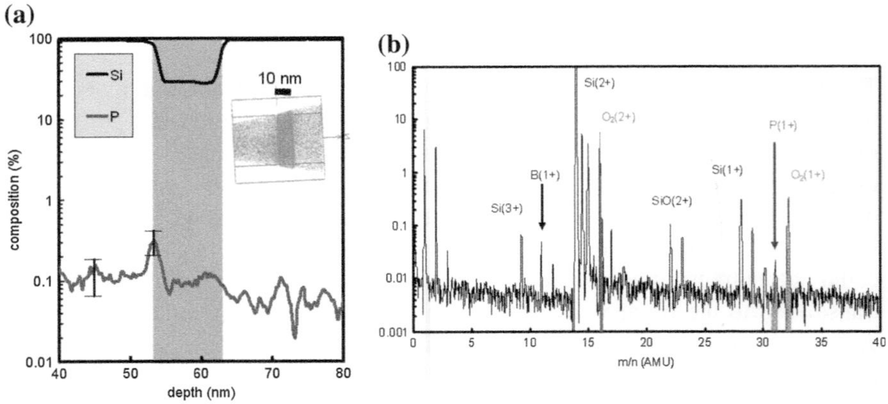

Fig. 5 **a** Concentration profile across a 10 nm SiO_2 layer in a silicon substrate (*inset*: reconstructed volume with SiO_2 layer represented in *gray*). Note the presence of a phosphorus segregation. **b** Mass spectrum related to the analysis. Note the presence of phosphorous and boron at a concentration of a few tens of ppm

mass resolution also leads to a much better sensitivity to low concentrations. The mass window to select elements can indeed be reduced so that the back ground noise is reduced proportionally to this width. The ultimate detection limit of LaWaTAP is a few tens of ppm (i.e. 0.001 at.%). One indirect advantage of laser pulses is that sample fractures are observed to be much less frequent, increasing therefore the depth that can be analyzed.

The physics of the interaction of ultra-fast laser pulses (a few hundreds fs) with the tip of whose radius is about 50 nm is complex and still under debate. The two main mechanisms that were involved in the evaporation process are based on a field effect (electric field of the electromagnetic field) or a thermal activation (Plasmon excitation and energy transfer to phonons) [9, 58]. According to the material and experimental conditions (intensity and duration (fs or ps laser) of the pulsed electric field), resonant photo-ionization can be observed. However, it is now accepted that even if the contribution of field enhancement and rectification of the laser field at the tip surface cannot be ruled out, thermal activation is the main process involved. However, due to the confinement of the field by the diffraction of the wave at the very end apex of the tip, this thermal process cannot be interpreted by conventional thermal cooling processes. An anomalous ultrafast cooling takes place on a time scale related to the confinement of the light absorption. Thus laser pulses lead to rapid heat pulses that promote field evaporation of surface atoms. Note that for materials of bad heat conductivity, the thermal pulse can last over a long time leading to a tailing edge of mass peaks. This deterioration of the mass spectra may be quite problematic for chemical species with close mass to charge (*m*/*n*) ratios. Larger trunk angle and shorter laser wavelengths generally improve the mass resolution [59, 60].

In order to assess the performance of the LaWaTAP, test samples were investigated and results were compared to depth profiles provided by SIMS

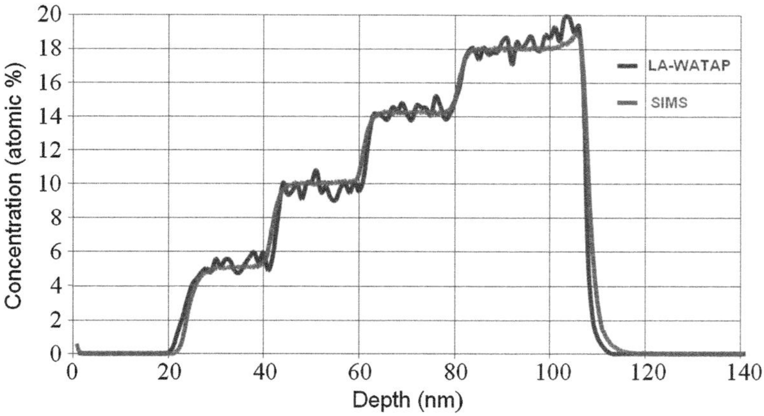

Fig. 6 Comparison of concentration profiles related to the analysis of a step-like SiGe multilayer (Maya) using LaWaTAP and SIMS (courtesy CAMECA). Reprinted figure with permission from Blavette et al. [12] by Carl Hanser Verlag GmbH

analysis. Test samples consisted in a stack of four thin SiGe layers (thickness close to 10 nm) with increasing Ge concentration (5, 10, 15, 18 at.% of Ge). Tips required for atom probe analysis were made using FIB annular milling [61]. The direct comparison of depth profiles provided by LaWaTAP and SIMS shows a very good agreement (Fig. 6). Compositions are observed to be quantitative and the depth scale is shown to be well calibrated. Similar comparisons on ultralow-energy As implants in silicon was made and also revealed good agreement between 3D atom-probe and SIMS [62].

APT measurements are however subjected to much larger fluctuations than SIMS as shown in Fig. 6. APT has the drawback of its advantage: due to the small analyzed area, profiles are shown to be subjected to larger statistical fluctuations. These larger sampling errors ($\sim 2\sigma$ with σ the standard deviation) are caused by the much smaller volume (i.e. smaller number (N) of atoms) on which atom-probe estimates relies on. In other words, APT has a higher spatial resolution and makes it possible to get concentrations on much smaller volumes. The obvious drawback is that sampling errors are larger.

The 3D atomic map of boron atoms in silicon samples containing boron deltas (ultra-thin boron-rich layers) is provided in Fig. 7. Four deltas appear in this reconstruction. The related depth profile taken perpendicular to boron layers is compared to SIMS profile in Fig. 8. Again a good agreement is observed. As expected, the boron layers are separated by 18 nm and a peak boron concentration around 10^{21} boron atoms/cm^3 ($\sim$ 2 at.%) is measured for each delta. Note a slow decrease of the peak concentration in the opposite direction of growth. Let us mention also the larger steepness of LaWaTAP profiles compared to SIMS. Interfaces appear more abrupt, illustrating therefore the higher depth resolution of LaWaTAP. However, the base level of boron between deltas is lower in SIMS profiles. LaWaTAP has a lower sensitivity compared to SIMS. As shown in Fig. 8,

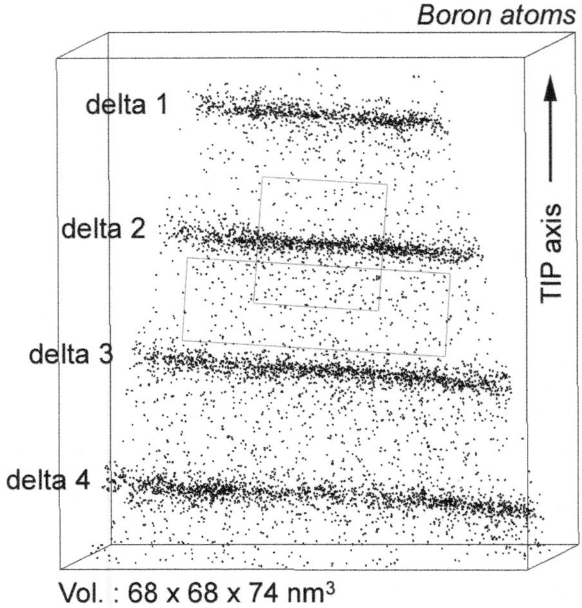

Fig. 7 3D reconstruction of boron distribution in silicon samples. For clarity, only boron atoms are showed. Four boron deltas are exhibited. Layers are separated by 18.3 nm. Courtesy Cadel et al. [63]. Reprinted with permission from Cadel et al. [63]. Copyright 2009, American Institute of Physics

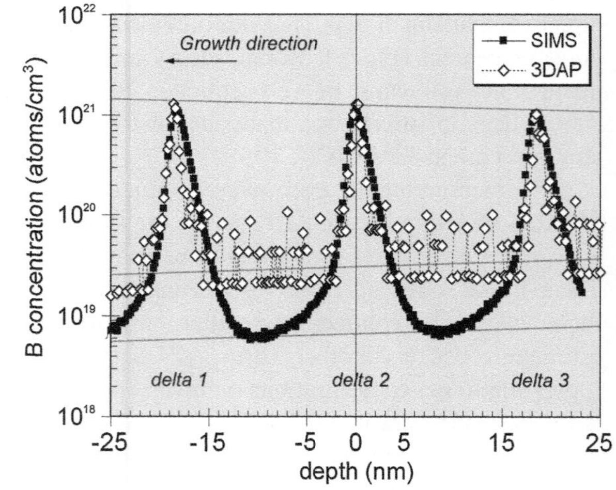

Fig. 8 Concentration profile of boron derived from 3D map (Fig. 7). A thin slice 0.2 nm thick was moved in a direction perpendicular to the three first boron deltas. The surface area of the sampling box is 25 × 25 nm. SIMS profile is superimposed to that of LaWaTAP. Reprinted with permission from Cadel et al. [63]. Copyright 2009, American Institute of Physics

the background in this particular experiment was larger than 10^{19} at/cm^3. A more detailed discussion is available in reference [63].

Laser assisted APT has shown to be a powerful approach for the investigation of dopant clustering in implanted silicon. An example relative to boron in implanted silicon is provided in Fig. 9. The high implantation dose (5×10^{15} at/cm^2, boron ions of 10 keV) was close to that of ultra-shallow junctions in last generation nanotransistors and exceeded the solubility limit of boron in Si. Samples were annealed at 600°C for 1 h. A native oxide (SiO_2—measured oxygen concentration:

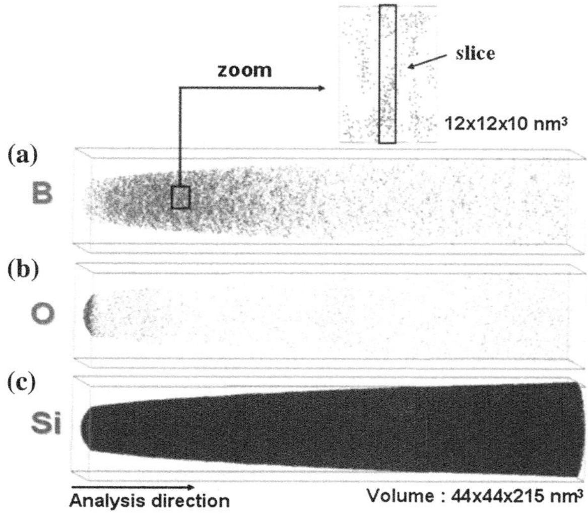

Fig. 9 Elemental maps of boron, oxide and silicon in the volume ($44 \times 44 \times 215$ nm^3) (boron implanted silicon annealed at 600°C for 1 h). A native oxide layer ($\approx$2 nm) is exhibited at the sample surface. The small slice used to construct concentration profile (Fig. 10) is represented in the zoom. Reprinted with permission from Cojocaru-Mirédin et al. [75]. Copyright 2009, Elsevier

63 at.%), 2 nm thick is observed at the surface. Note the drastic decrease of the boron concentration for depth exceeding 100 nm. Numerous platelet-shaped boron clusters containing about fifty atoms (Si and B) were discovered (zoom region). As expected, the number density is significantly higher in the region where the boron concentration is higher (higher driving force for clustering). The number density of clusters ($\sim 10^{18}$ clusters/cm^3) is observed to decrease drastically for depth exceeding 100 nm. These clusters have the shape of platelets parallel to the implanted surface, in agreement with BICs (boron interstitial clusters) observed by Cristiano et al. [64]. The average boron content in these cluster as determined from LaWaTAP images was found close to 7 at.% [65].

Related concentration profile of boron (in log scale) as derived from Fig. 9 is compared to SIMS profile in Fig. 10. This profile was taken in the Ga-free region of the reconstructed region in order to minimise Ga-irradiation effects. A fairly good agreement between SIMS and LaWaTAP depth profiles is observed. The maximum concentration (9×10^{20} at/cm^3), close for both techniques, is detected at a depth of 35 nm in both APT and SIMS profiles (implantation peak). Significant differences are however apparent. Again, statistical fluctuations appear in the LaWaTAP depth profile. The atom-probe profile gives a boron concentration of 2×10^{20} at/cm^3 close to the sample surface whereas the SIMS profile shows a boron concentration twice larger (4.2×10^{20} at/cm^3). It is thought that SIMS overestimate boron concentration close to the surface. SIMS is probably less quantitative at the beginning of analysis close to the surface. More, the presence of the native oxide might also have an influence [66].

It is worth mentioning that the broadly Gaussian implantation profile of boron (Fig. 10) leads to a non constant boron supersaturation. Therefore the driving force for nucleation and as a result the number density of clusters varies along the depth in a similar way as the depth profile of boron [67].

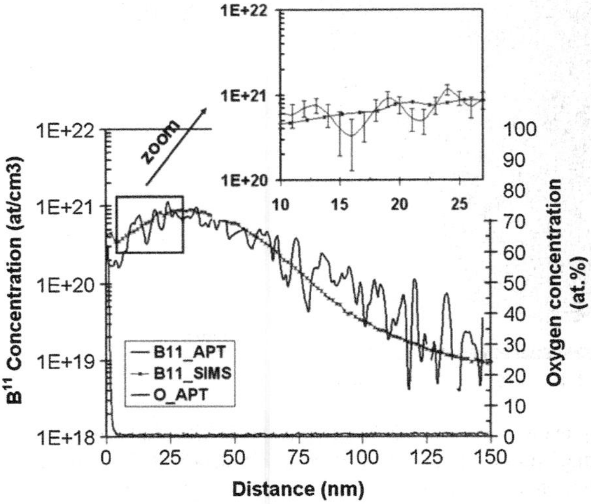

Fig. 10 B^{11} concentration profile as given by SIMS and LaWaTAP (APT: atom probe tomography). LaWaTAP profile reveals the presence of a native oxide at the sample surface. LaWaTAP depth profiles were obtained by moving a small box (2 nm thick and 12 nm wide) through the analyzed volume (Fig. 9) in a direction perpendicular to the sample surface that was exposed to boron implantation. Reprinted with permission from Cojocaru-Miredin et al. [65]. Copyright 2009, Elsevier

Laser assisted APT is also a powerful approach for the investigation of tunnel junctions used in non volatile storage computer devices and in sensor applications. Theoretically, based on ab inito calculations, a considerable change in the resistance of up to 1,000% was predicted for a double stack of Fe/MgO/Fe tunnel junctions. Unfortunately this was not yet verified experimentally. The origin of this lower TMR performance could come from interfaces. It is therefore of utmost importance to investigate the distribution of chemical species close to interfaces. The atomic map related to the investigation of Fe/MgO/Fe tunnel junction is provided in Fig. 11. Layers were grown on a highly doped flat Si-(100) substrate and subsequently prepared in the FIB.

APT images clearly reveal the Fe/MgO/Fe tunnel junction with well defined interfaces. The thickness of the MgO layer is found close to 4 nm as expected. A close examination of the image however reveals the presence of oxygen in the second Fe layer (left hand). This is confirmed in composition profiles (Fig. 11d) that were drawn perpendicularly to interfaces. A remarkable result is the asymmetry of the oxide on both sides of the MgO barrier. Profiles clearly indicate that the second MgO/Fe interface (2) is more diffuse than the first interface (1). The first interface (1) is dominated by a clear drop of the Fe composition and a steep increase in the Mg and O composition within less than 1 nm. This interface shows barely the formation of iron oxides. In contrast, the second interface (2) is mainly characterized by the formation of FeO or Fe_2O oxides. The formation of such oxides close to the MgO/Fe interface could result in a degradation of the TMR-effect [12].

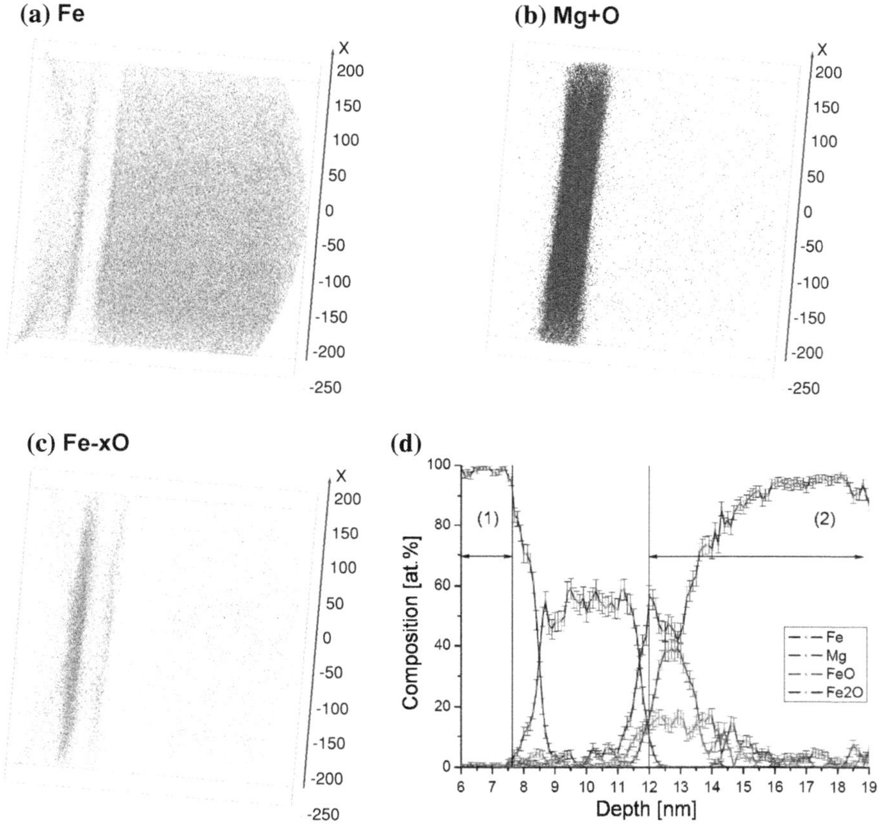

Fig. 11 3D reconstruction of a Fe/MgO/Fe tunnel junction. The MgO (4 nm) oxide was deposited on a Fe layer (**a**). The thicker Fe layer was then deposited on MgO (**b**). Interfacial segregation of oxygen (Fe-xO) at the Fe·MgO interface (**c**). Depth profiles show that oxygen has diffused in the Fe layer on which the MgO layer was deposited. Courtesy of Gilbert and Al Kassab [12]). Composition profile through the respective barrier-layer showing the distribution of the different elements on both sides of the metal/oxide/metal interfaces (**d**)

5 Conclusions

Because of its very high spatial resolution and its quantitativity, APT has played a major role in physical metallurgy in particular for the investigation of the early stages of phase separation and impurity segregation to lattice defects. Recent developments in transmission electron microscopy has been very impressive and demonstrate that electron tomography at a comparable scale is now available. However the atomic-scale is not yet achieved in electron tomography and quantitative data on the composition of nano-objects appear to be less evident or straightforward than with APT. Comparison between reconstructed volumes were recently achieved [68]. Electron energy loss spectroscopy, energy filtering as well as energy dispersive techniques are alternative nanoanalysis techniques that can be combined with APT in a very fruitful way. Because of the rather small volume that can be observed with APT, the instrument requires the combination of other microscopic approaches able to get both a larger and more representative view of the material microstructure.

Since the emergence of the first generations in the early nineties, major improvements have been brought about to APT. The much larger area of analysis available (30 times than first generations, $\sim$ 100 nm), the higher analysis rates (10^7 ions/h) and the implementation of ultrafast laser have been clearly decisive breakthroughs. They provided a new impulse to 3D atom-probe in material science pushing the instrument out of the confidentiality that it knew before. FIB ion milling also contributed in a major way to open APT more widely to Nanosciences (spin valves, nano-wires, nano-powders, ultra-shallow junctions in microelectronics, MOS-FET nanotransistors). Laser has also provided the apparatus with a higher mass resolution and a better sensitivity.

The use of laser pulses instead of electric pulses had incidentally another advantage. It enables the use of a lower DC field (V_0) that reduces the electrostatic stress and avoid cyclic stress induced by voltage pulses. This considerably makes it easier the investigation of brittle metallic alloys (irradiated steels, oxides, titanium, intermetallics, carbides...) that were often subjected to frequent rupture under the cyclic electrostatic stress applied and increased life time of tips.

The SIMS is the traditional instrument used in microelectronics to get implantation profiles. However, this instrument approaches its ultimate limits for the investigation of last generations of MOS-FET transistors the dimensions of which are below 100 nm. The laser assisted 3D atom-probe should play an increasing role in this field. This instrument has many advantage compared to SIMS (3D imaging, spatial resolution) but it has also several shortcomings or drawbacks (preparation of specimen, mass resolution, statistics and sensitivity). Instruments are in fact less concurrent than complementary. Statistical fluctuations in LaWaTAP are larger because of the smaller volume analyzed compared to SIMS (50 μm in diameter as compared to 50 nm for LaWaTAP). The analyzed area of SIMS, around 10^6 larger, leads to a much better sensitivity. Even if the low ionisation efficiency of SIMS (between 0.1 and 1%) compared to LaWaTAP (ionization rate = 1, detection efficiency $Q = 0.5$) reduces the difference of collected ions (N), statistics remain much better in SIMS analysis (a factor close to 30). Using electrostatic lens or energy compensation systems (electrostatic mirror), the ultimate sensitivity of the instrument reaches 10 ppm. However, this advantage is counterbalanced by the higher lateral resolution of LaWaTAP and its 3D imaging capability.

The volumes that are reconstructed with APT have now a size close to that of last generation MOS nanotransistors [69]. Imaging the dopant distribution in 3D at the atomic scale within nano-transistors is now accessible. Making such experiments in real integrated circuits using FIB and lift-out techniques remains however a difficult task and a challenge for future. Impressive results were recently published both on this topics [70]. Recent papers give more details on experimental conditions and potential of this new generation of atom-probe in nanosciences [71–74].

Acknowledgments Many thanks to CAMECA for its financial support and for many fuitfull discussions. L. Renaud (CAMECA-France) is acknowledged for providing results on SiGe multilayers. Parts of this work were conducted within programs supported by the ANR Pnano call, the ESP CARNOT Institut. The authors wish to thank D. Mangelinck (IN2MP, Marseille) for the many fruitful discussions we have had on common researches we made on boron implants in silicon. Thanks to O. Cojocaru, M. Gilbert, T. Al Kassab, E. Cadel for providing results presented in this review.

References

1. Müller, E.W., Panitz, J., Mc Lane, S.B.: Atom-probe field ion microscope. Rev. Sci. Instrum. **39**, 83 (1968)
2. Cerezo, A., Godfrey, I.J., W Smith, G.D.: Application of a position-sensitive detector to atom probe microanalysis. Rev. Sci. Instrum. **59**, 862–866 (1988)
3. Blavette, D., Bostel, A., Sarrau, J.M., Deconihout, B., Menand, A.: An atom-probe for 3-dimensional tomography. Nature **363**, 432–435 (1993)
4. Vurpillot, F., Da Costa, G., Menand, A., Blavette, D.: Structural analyses in three-dimensional atom probe: a Fourier approach. J. Microsc. **203**, 295–302 (2001)
5. Gault, B., Moody, M.P., De Geuser, F., La Fontaine, A., Stephenson, L.T., Haley, D., Ringer, S.P.: Spatial resolution in atom probe tomography. Microsc. Microanal. **16**, 99–110 (2010)
6. Vurpillot, F., Renaud, L., Blavette, D.: A new step towards the lattice reconstruction in 3DAP. Ultramicroscopy **95**, 223–229 (2003)
7. Cottrell, A.H., Bilby, B.A.: Distribution of solute atoms round a slow dislocation. Proc. Phys. Soc. Lond. **A62**, 49 (1949)
8. Blavette, D., Cadel, E., Fraczkiewicz, A., Menand, A.: Three-dimensional atomic-scale imaging of impurity segregation to line defects. Science **17**, 2317 (1999)
9. Gault, B., Vella, A., Vurpillot, F., Menand, A., Blavette, D., Deconihout, B.: Optical and thermal processes involved in ultrafast laser pulse interaction with a field emitter. Ultramicroscopy **107**, 713 (2007)
10. Gault, B., Vurpillot, F., Vella, A., Gilbert, M., Menand, A., Blavette, B., Deconihout, B.: Design of a femtosecond laser assisted tomographic atom probe. Rev. Sci. Instr. **77**, 043705 (2006)
11. Thompson, K., Flaitz, P.L., Ronsheim, P., Larson, D.J., Kelly, T.F.: Imaging of arsenic Cottrell atmospheres around silicon defects by three-dimensional atom probe tomography. Science **317**, 1370 (2007)
12. Blavette, D., Al Kassab, T., Cadel, E., Mackel, A., Gilbert, M., Cojocaru, O., Deconihout, B.: Laser-assisted atom probe tomography and nanosciences. Intern. J. Mater. Res. **99**, 454 (2008)
13. Miller, M.K., Russell, K.F., Thompson, K., Alvis, R., Larson, D.J., Miller, M.K.: Review of atom probe FIB-based specimen preparation methods. Microsc. Microanal. **13**, 428–436 (2007)
14. Müller, E.W.: Das Feldionenmikroskop. Zeitschrift Fur Physik **131**, 136–142 (1951)
15. Blavette, D., Deconihout, B., Bostel, A., Sarrau, J.M., Bouet, M., Menand, A.: The tomographic atom-probe: a quantitative 3D nanoanalytical instrument on an atomic-scale. Rev. Sci. Instrum. **64**, 2911–2919 (1993)
16. Deconihout, B., Renaud, L., Bouet, M., Da Costa, G., Bostel, A., Blavette, D.: Implementation of the optical tap: preliminary results. Ultramicroscopy **73**, 253–260 (1998)

17. Da Costa, G., Vurpillot, F., Bostel, A., Bouet, M., Deconihout, B.: Design of a delay-line position-sensitive detector with improved performance. Rev. Sci. Instrum. **76**, 013304 (2005)
18. Bas, P., Bostel, A., Deconihout, B., Blavette, D.: A general protocol for the reconstruction of 3D atom probe data. Appl. Surf. Sci. 87/88, 298–304 (1995)
19. Vurpillot, F., Bostel, A., Menand, A., Blavette, D.: Trajectories of field emitted ions in 3D atom probe. Europ. Phys. J. Appl. Phys. **6**, 217–221 (1999)
20. Vurpillot, F., Bostel, A., Blavette, D.: The shape of field emitters and the ion trajectories in 3D atom probe. J. Microsc. Oxford **196**, 332–336 (1999)
21. Vurpillot, F., Bostel, A., Cadel, E., Blavette, D.: The spatial resolution of 3D atom probe in the investigation of single-phase materials. Ultramicroscopy **84**, 213–224 (2000)
22. Vurpillot, F., Bostel, A., Blavette, D.: Trajectory overlaps and local magnification in 3D atom probe. Appl. Phys. Lett. **76**, 3127–3129 (2000)
23. Vurpillot, F., Cerezo, A., Blavette, D., Larson, D.J.: Modelling image distortions in 3DAP. Microsc. Microanal. **10**, 384–390 (2004)
24. De Geuser, F., Lefebvre, W., Danoix, F., Vurpillot, F., Forbord, B., Blavette, D.: An improved reconstruction procedure for the correction of local magnification effects in 3D atom probe. Surf. Inter. Anal. **39**, 268–272 (2007)
25. Philippe, T., Gruber, M., Vurpillot, F., Blavette, D.: Clustering and local magnification effects in atom probe tomography: a statistical approach. Microsc. Microanal. **16**(5), 643–648 (2010)
26. Hetherington, M.G., Miller, M.K.: Some aspects of the measurement of composition in the atom probe. J. De Phys. **50**, 535–540 (1989)
27. Hyde, J.M., English, C.A.: Microstructual processes in irradiated materials. In: Lucas, R.G.E., Snead, L., Kirk, M.A.J., Elliman, R.G. (eds.) MRS 2000 Fall Meeting Symposium, Boston, MA, pp. 27–29 (2001)
28. Vaumousse, D., Cerezo, A., Warren, P.J.: A procedure for quantification of precipitate microstructures from three-dimensional atom probe data. Ultramicroscopy **95**, 215–221 (2003)
29. Pearson, K.: On the Theory of Contingency and Its Relation to Association and Normal Correlation, Biometric Series No. 1. Drapers' Co. Memoirs, London (1904)
30. Moody, M.P., Stephenson, L.T., Liddicoat, P.V., Ringer, S.P.: Contingency table techniques for three dimensional atom probe technology. Microsc. Res. Tech. **70**, 258–268 (2007)
31. Hellman, O.C., Vandenbroucke, J.A., Rüsing, J., Isheim, D., Seidman, D.N.: Analysis of three-dimensional atom-probe data by the proximity histogram. Microsc. Microanal. **6**, 437–444 (2000)
32. Geiser, B.P., Kelly, T.F., Larson, D.J., Schneir, J., Roberts, J.: Spatial distribution maps for atom probe tomography. Microsc. Microanal. **13**, 437–447 (2007)
33. Vurpillot, F., De Geuser, F., Blavette, D.: Application of the Fourier transform and autocorrelation to cluster identification in 3DAP. J. Microsc. **216**, 234–240 (2004)
34. Marteau, L., Pareige, C., Blavette, D.: Imaging the three orientation variants of the Do22 phase by 3DAP microscopy. J. Microsc. **204**, 247–251 (2001)
35. Philippe, T., De Geuser, F., Duguay, S., Lefebvre, W., Cojocaru-Mirédin, O., Da Costa, G., Blavette, D.: Clustering and nearest neighbour distances in atom probe tomography. Ultramicroscopy **109**, 1304 (2009)
36. Stephenson, L.T., Moody, M.P., Liddicoat, P.V., Ringer, S.P.: New techniques for the analysis of fine scaled clustering phenomena within atom probe tomography data. Microsc. Microanal. **13**, 448–463 (2007)
37. De Geuser, F., Lefebvre, W., Blavette, D.: 3D atom probe investigation of solute atoms clustering during natural ageing or preageing in an almgsi alloy. Phil. Mag. **86**, 227–234 (2006)
38. Philippe, T., Duguay, S., Blavette, D.: Clustering and pair correlation function in atom probe tomography. Ultramicroscopy **109**, 1304–1309 (2010)
39. Pareige, P., Auger, P., Bas, P., Blavette, D.: Direct observation of precipitation in neutron irradiated FeCu alloys by atomic tomography. Scripta. Met. **33**, 1033–1036 (1995)

40. Cadel, E., Frackiewicz, A., Blavette, D.: Suzuki effect in boron-doped feal intermetallics. Scripta. Mater. **51**, 437–441 (2004)
41. Cadel, E., Lemarchand, D., Chambreland, S., Blavette, D.: Atom probe tomography investigation of the microstructure of superalloys N18. Acta. Mater. **50**, 957–966 (2002)
42. Schmuck, C., Caron, P., Hauet, A., Blavette, D.: Ordering and precipitation in low supersaturated nicral model alloy: an atomic scale investigation. Phil. Mag. A **76**, 527–542 (1997)
43. Amouyal, Y., Mao, Z., Seidman, D.N.: Phase partitioning and site-preference of hafnium in the gamma '(L1(2))/gamma(fcc) system in Ni-based superalloys: an atom-probe tomographic and first-principles study. Appl. Phys. Lett. 95, 161909 (2009)
44. Pareige, C., Soisson, F., Martin, G., Blavette, D.: Ordering and phase separation in nicral alloys: Monte Carlo simulation and 3D atom probe study. Acta. Met. Mater. **47**, 1889–1899 (1999)
45. Danoix, F., Auger, P., Blavette, D.: Hardening of aged stainless steels by spinodal decomposition. Microsc. Microanal. **10**, 03 (2004)
46. Blavette, D., Cadel, E., Pareige, C., Deconihout, B., Caron, P.: Phase transformation and segregation to lattice defects in Ni-base superalloys. Microsc. Microanal. **13**, 1–20 (2007)
47. Danoix, F., Grancher, G., Bostel, A., Blavette, D.: Standard deviation of composition measurements in atom probe analyses: Part I conventional 1D atom probe. Ultramicroscopy **107**, 734–739 (2007)
48. Danoix, F., Grancher, G., Bostel, A., Blavette, D.: Standard deviation of composition measurements in atom probe analyses: Part II 3DAP. Ultramicroscopy **107**, 739–743 (2007)
49. Miller, M.K., Hetherington, M.G.: Local magnification in the atom probe. Surf. Sci. **246**, 443–449 (1991)
50. Blavette, D., Vurpillot, F., Pareige, P., Menand, A.: A model accounting for the spatial overlaps of 3DAP's. Ultramicroscopy **89**, 145–153 (2001)
51. Deschamps, A., Bigot, A., Auger, P., Brechet, Y., Livet, F., Blavette, D.: A comparative study of precipitate composition in Al-Zn-Mg using tomographic atom probe and Saxs. Phil. Mag. A **81**, 2391–2414 (2001)
52. Lettelier, L., Guttmann, M., Blavette, D.: Atomic scale investigation of grain-boundary microchemistry in boron-doped nickel-base superalloys astroloy with a 3D atom-probe. Phil. Mag. **70**, 189–194 (1994)
53. Blavette, D., Duval, P., Letellier, L., Guttmannn, M.: Atomic-scale Apfim and Tem investigation of Gb microchemistry in Astroloy nickel base superalloys. Acta. Met. Mater. **44**, 4995–5005 (1996)
54. Lemarchand, D., Cadel, E., Chambreland, S., Blavette, D.: Investigation of grain boundary structure-segregation relationship in a N18 nickel-based superalloy. Phil. Mag. A **82**, 1651–1669 (2002)
55. Kellogg, G., Tsong, T.T.: Pulsed laser atom-probe. J. Appl. Phys. (USA) 51, 1184 (1980)
56. Deconihout, B., Vurpillot, F., Gault, B., Da Costa, G., Bouet, M., Bostel, A., Blavette, D., Hideur, A., Martel, G., Brunel, M.: Towards a laser assisted wide angle tomographic atom probe. Surf. Interface Anal. **39**, 278 (2007)
57. Kinkgham, D.R.: The post-ionization of field evaporated ions: a theoretical explanation of multiple charge states. Surf. Sci. **116**, 273–301 (1982)
58. Vella, A., Deconihout, B., Marrucci, L., Santamato, E.: Femtosecond field ion emission by surface optical rectification. Phys. Rev. Lett. **99**, 046103 (2007)
59. Houard, J., Vella, A., Vurpillot, F., Deconihout, B.: Optical near-field absorption at a metal tip far from plasmonic resonance. Phys. Rev. B **81**, 125411 (2010)
60. Bunton, J.H., Olson, J.D., Lenz, D.R., Kelly, T.F.: Advances in pulsed-laser atom probe: instrument and specimen design for optimum performance. Microsc. Microanal. **13**, 418–427 (2007)
61. Larson, D.J., Foord, D.T., Petford-Long, A.K., Liew, H., Blamire, M.G., Cerezo, A., Smith, G.D.W.: Field-ion specimen preparation using focused ion-beam milling. Ultramicroscopy **79**, 287 (1999)

62. Thompson, K., Bunton, J.H., Kelly, T.F., Larson, D.J.: Characterization of ultralow-energy implants and towards the analysis of three-dimensional dopant distributions using three-dimensional atom-probe tomography. J. Vac. Sci. Technol. B **24**, 412 (2006)
63. Cadel, E., Vurpillot, F., Deconihout, B.: Depth resolution function of the laser assisted tomographic atom probe in the investigation of semiconductors. J. Appl. Phys. **106**, 044908 (2009)
64. Cristiano, F., Hebras, X., Cherkashin, N., Claverie, A.: Clusters formation in ultralow-energy high-dose boron-implanted silicon. Appl. Phys. Lett. **83**, 5407 (2003)
65. Cojocaru-Miredin, O., Cadel, E., Vurpillot, F., Mangelinck, D., Blavette, D.: Three-dimensional atomic-scale imaging of boron clusters in implanted silicon. Scripta. Mater. **60**, 285 (2009)
66. Gurenko, A.A., Veksler, I.V., Meixner, A., Thomas, R., Dorfman, A.M., Dingwell, D.B.: Matrix effect and partitioning of boron isotopes between immiscible Si-rich and B-rich liquids in the Si-Al-B-Ca-Na-O system: a SIMS study of glasses quenched from centrifuge experiments. Chem. Geol. **222**, 268 (2005)
67. Cojocaru-Mirédin, O., Mangelinck, D., Blavette, D.: Nucleation of boron clusters in implanted silicon. J. Appl. Phys. **106**, 113525 (2009)
68. Arslan, I., Marquis, E.A., Homer, M.: Towards better 3-D reconstructions by combining electron tomography and atom-probe tomography. Ultramicroscopy **108**, 1579–1585 (2008)
69. Kelly, T.F., Miller, M.K.: Invited review article: atom probe tomography. Rev. Sci. Instrum. 78, 031101 (2007)
70. Inoue, K., Yano, F., Nishioda, A., Takamizawa, H., Nagai, Y., Hasegawa, M.: Dopant distributions in n-MOSFET structure observed by atom probe tomography. Ultramicroscopy **109**, 1479–1484 (2009)
71. Sha, G., Ringer, S.P.: Effect of laser pulsing on the composition measurement of an Al-Mg-Si-Cu alloy using three-dimensional atom probe. Ultramicroscopy **109**, 580 (2009)
72. Torres, K.L., Thompson, G.B.: Grain boundary enrichment in the FePt polymorphic A1 to L1(0) phase transformation. Ultramicroscopy 109, 606 (2009)
73. Gruber, M., Oberdorfer, C., Stender, P., Shmitz, G.: Laser-assisted atom probe analysis of sol-gel silica layers. Ultramicroscopy **109**, 654 (2009)
74. Nishimura, A., Nogiwa, K., Otobe, T., Okhubo, T., Hono, K., Kondo, K., Yokoyama, A.: Influence of laser irradiation condition on a femtosecond laser-assisted tomographic atom probe. Ultramicroscopy **109**, 467 (2009)
75. Cojocaru-Mirédin, O., Cadel, E., Blavette, D., Mangelinck, D., Hoummada, K., Genevois, C., Deconihout, B.: Atomic-scale redistribution of Pt during reactive diffusion in Ni(Pt)-Si contacts. Ultramicroscopy **109**, 797–801 (2009)

GPSR Compliance

The European Union's (EU) General Product Safety Regulation (GPSR) is a set of rules that requires consumer products to be safe and our obligations to ensure this.

If you have any concerns about our products, you can contact us on ProductSafety@springernature.com

In case Publisher is established outside the EU, the EU authorized representative is:

Springer Nature Customer Service Center GmbH
Europaplatz 3
69115 Heidelberg, Germany

Batch number: 10614940

Printed by Printforce, the Netherlands